Palgrave Studies in Economic History

Series Editor
Kent Deng, London School of Economics, London, UK

Palgrave Studies in Economic History is designed to illuminate and enrich our understanding of economies and economic phenomena of the past. The series covers a vast range of topics including financial history, labour history, development economics, commercialisation, urbanisation, industrialisation, modernisation, globalisation, and changes in world economic orders.

Antonie Doležalová · Catherine Albrecht
Editors

Behind the Iron Curtain

Economic Historians During the Cold War, 1945–1989

palgrave
macmillan

Editors
Antonie Doležalová
Institute of History
Czech Academy of Sciences
and Faculty of Social Sciences
Charles University
Prague, Czech Republic

Catherine Albrecht
Ohio Northern University
Ada, OH, USA

ISSN 2662-6497 ISSN 2662-6500 (electronic)
Palgrave Studies in Economic History
ISBN 978-3-031-31580-0 ISBN 978-3-031-31578-7 (eBook)
https://doi.org/10.1007/978-3-031-31578-7

Cover illustration: Cum Okolo/Alamy Stock Photo

This Palgrave Macmillan imprint is published by the registered company Springer Nature Switzerland AG
The registered company address is: Gewerbestrasse 11, 6330 Cham, Switzerland

This book is dedicated to the memory of scientists from around the world
who due to the ideological or geopolitical obstacles
have not been allowed to experience
the joy of free exchange
of scientific ideas.

Acknowledgments

The Iron Curtain's fall in autumn 1989 announced the end of the Cold War. It seemed that the world had been changed forever and even that history ended. The early morning of February 24, 2022, was for many people an unwanted *déjà vu*. The time they wanted to forget was back. That early morning brutally revealed that history did not end in 1989 and demonstrated how easy it is to divide countries and their people into "we" and "they." Significant obstacles started to be erected in Europe not only for greedy businessmen, but also for ordinary people, artists, and scientists. A new generation of scientists had to face the same threats, disappointments, and disillusions as did the generations that lived and worked under the pressure of the Cold War; only the border of the West moved to the East.

From the very beginning, this book was intended to be a contribution to the discussion about the principles of free academic cooperation despite the geopolitical and ideological barriers as there still exist countries in the world today in which study and research are not free intellectual activities. At the same time, this book also sought to commemorate the scientists who found the courage to break through the ideological and political obstacles they faced in their academic work during the Cold War era.

The idea of writing the book has a six-year-long history. It came into being in a sunny morning in June 2016 in Bratislava, when my Slovak colleague, Roman Holec, and I discussed ideas to be proposed for a panel for the World Economic History Congress in Boston in 2018. The session proposal was inspired by the fiftieth anniversary of the last WEHC in the United States and by the fact that it was that congress in 1968 that made the future prominent economic historian, Alice Teichova, to leave Czechoslovakia together with her husband, Mikuláš Teich, just a few days after the Warsaw Pact invasion of Czechoslovakia. Two years later, they decided for emigration. The story of their lives and careers demonstrates the complicated relationship between home and exile as well as between Eastern and Western historiography.

Two years later, three of the authors of this book participated in a panel at Boston with Catherine Albrecht as a discussant. Here also Palgrave approached Catherine Albrecht and me to publish a book based on the panel. Since then, I have been developing the idea of the book project through networking to gather authors from all post-communist European countries; the book contains chapters on seven of eight of those countries.

Thus, my thanks goes to eight brave men and one woman, all excellent researchers, who accepted my invitation to write on the obstacles and difficulties their predecessors on the eastern side of the Iron Curtain had to face. I want to thank them for their patience and openness to include further and further aspects into their chapters as the project developed and new perspectives were revealed.

I am thankful to Professor Catherine Albrecht that she embarked that journey with us, commented on our texts in Boston, and transferred here and there Central European English into American phrasing.

I owe my thanks to the Ministry of Education of the Czech Republic, which financially supported the gathering of the team and the initial archive research; to Charles University and its Faculty of Social Sciences, which supported editing of the book; and to the National Scholarship Program of the Slovak Republic for its generous support thanks to which I could realize the final stage of the research and join Roman Holec to write the Czechoslovak chapter in Bratislava in the spring of 2021.

Last but not least, I want to express my gratitude to Mikuláš Teich and Miroslav Hroch, two outstanding and world-renown historians who experienced the evil of the divided world during the Cold War era and strove to overcome the obstacles introduced by the governing regime in Czechoslovakia. I will always be grateful for their willingness to share with me their recollections, sometimes bitter, and reflections on how the historiography, opportunities for cooperation, and modes of transfer of ideas developed during the Cold War era and since that time.

An attentive reader will recognize how much the past is still part of our lives, relations, biases, and everyday routines. The book demonstrates that for scientists who pursued truth the Iron Curtain was a state of mind. In pursuit of truth, the scientific community must share its findings and defend its freedom. The question only is, at what price.

Spring 2022

Antonie Doležalová
Institute of History
Czech Academy of Sciences
and Faculty of Social Sciences
Charles University
Prague, Czech Republic

Contents

Editors and Contributors

About the Editors

Antonie Doležalová is Associate Professor in Economic History at the Faculty of Social Sciences at Charles University and Research Fellow in the Institute of History of the Czech Academy of Sciences, both in Prague, and Senior Member at Robinson College in Cambridge. In her research, she focuses on economic history, business history, historiography of economic history, and history of economic thought in Central and Eastern Europe in the nineteenth and twentieth centuries.

Catherine Albrecht is Professor of History Emerita at Ohio Northern University in Ada, Ohio. She is an expert in the history of economic thought in East Central Europe, having published numerous articles on the intersection of economic development and nationalist competition in the late nineteenth and early twentieth centuries.

Contributors

Axel Fair-Schulz is Associate Professor of Modern European History at the State University of New York at Potsdam. His research focuses on the intersections of intellectual, economic, and political history as well as the transfer of ideas across different political cultures and ideological systems, including trans-Atlantic exchanges.

Matei Gheboianu is Associate Professor at the Faculty of History, University of Bucharest. His main research fields are Romanian higher education during communism and mass media in post-communism.

Roman Holec is Professor of History at the Faculty of Art of Comenius University and Research Fellow in the Institute of History of the Slovak Academy of Sciences, both in Bratislava (Slovakia). Areas of his scientific interest are economic and social history, environmental history, and the history of aristocracy, all from 1848 to 1945 in Central Europe.

György Kövér is Professor Emeritus at the Department of Economic and Social History at Eötvös Loránd University in Budapest, member of the Hungarian Academy of Sciences, and President of the Hungarian National Committee of the International Economic History Association. He publishes in the field of economic history and historiography.

Volodymyr V. Kravchenko is Professor of History at the Department of History, Classics, and Religion at the Faculty of Arts and Director of the Contemporary Ukraine Studies Program at the Canadian Institute of Ukrainian Studies, both at the University of Alberta. His research interests cover modern historiography, Ukrainian and Russian history of the nineteenth and twentieth centuries, and Kharkiv city and region.

Žarko Lazarevič is Professor of History and Head of the Department of Economic and Social History at the Institute of Contemporary History in Ljubljana. He focuses on the development of agriculture, the industrialization process, cooperatives, entrepreneurialism, and the role of nationalism in the economy in Slovenia within the European context in the late nineteenth and twentieth centuries.

Ionuţ-Mircea Marcu is a Ph.D. candidate in History at the University of Bucharest and École hautes études en sciences sociales in Paris. He works on Romanian history writing in late socialism and early post-communism.

Bogdan Murgescu is Professor in History and Vice-Rector at the University of Bucharest. He served two terms as President of the Romanian Society for Historical Sciences (2011–19) and has been elected member of Academia Europea (2020).

Anna Sosnowska is Associate Professor of Sociology at the University of Warsaw. Her research focuses on the historical sociology of Polish and Jewish migrations to the United States and on economic development in Eastern Europe.

CHAPTER 1

Introduction: Economic History Revisited

Antonie Doležalová

In 1977, Alice Teichova (1920–2015) and Mikuláš Teich (1918–2018), two Czechoslovak historians in British exile, attended the International Natural Sciences Congress in Edinburgh. For the first time since their emigration to the UK after the 1968 Warsaw Pact invasion of Czechoslovakia, they had an opportunity to meet their former Czechoslovak colleagues who also attended the congress. However, during the congress opening reception, the Czechoslovaks acted as though they did not see them. When Mikuláš asked the leader of the Czechoslovak delegation, Jaroslav Purš, why they did not talk to them, Purš responded: Mikuláš, we stand at the opposite side of a barricade.

In the most common historical narrative, the Iron Curtain is still considered a barricade. A barricade is used for fighting, not for cooperating. For forty years of the twentieth century, the Iron Curtain divided Western and Eastern Europe. Through propaganda's glasses and disinformations, East Europeans saw West Europeans as reactionary elements and avaricious imperialists; West Europeans demonized East Europeans

A. Doležalová (✉)
Institute of History, Czech Academy of Sciences and Faculty of Social Sciences, Charles University, Prague, Czech Republic
e-mail: ad875@cam.ac.uk

A. Dolezalova and C. Albrecht (eds.), *Behind the Iron Curtain*, Palgrave Studies in Economic History,
https://doi.org/10.1007/978-3-031-31578-7_1

as thickheaded crusaders for a utopian communist world willing to shoot whenever they saw the slightest attempt of someone trying to cross the border of the socialist camp from either side. Both West and East Europeans thought of themselves as the sole rightful and enlightened defenders of freedom and peace.

The Aim of the Book: To Revise the Iron Curtain

Using the example of economic history, this book challenges the narrative of the non-cooperative nature of scientific work during the Cold War due to socialist scientists' incapability and disinclination to engage openly in international discussions. It introduces famous as well as almost forgotten scholars from former socialist countries who attempted to eliminate the Iron Curtain and strove to break through obstacles to freedom in the transfer of scientific ideas. In doing so, the book casts new light on the conditions under which scientists had to work in socialist countries and on the flow of ideas between the East and West during the Cold War from the perspective of the eastern side of the Iron Curtain.

In global historiography, significant streams of contemporary intellectual history are analyzing the transfer of ideas and cooperation among the scientists on both sides of the Iron Curtain. This book benefits from international discussion in economic history that incorporates a comparative analysis of the entire discipline (Hudson & Boldizzoni, 2016), the history of the professional associations that were established to create a platform for East-West discussions and meetings (Berg, 2015), the Cold War studies that offer an analytical tool for inquiry into the scientific work during the Cold War period in a global perspective (van Dongen, 2015), as well as from the comparison of different academic cultures on both sides of the Iron Curtain (Gienow-Hecht & Donfried, 2010) and the experience of Central and East European scientists in Western exile (Zadencka et al., 2015). All these aspects are the subject of interest in this book. Its novelty, however, is the until-now unrecognized changes in institutional conditions for international cooperation in individual socialist countries throughout the Cold War era that defined the possibilities for East European scientists to participate in international cooperation and influenced their motivations and decision-making process as they considered whether to cooperate with their Western peers.

Initially, the book project considered the hypotheses that governments on the east side of the Iron Curtain, being orchestrated by the USSR, obstructed East-West cooperation in the same way in all countries and that due to the multiple obstacles, cooperation, and the free flow of ideas and innovations between the Eastern and Western scientists during the Cold War era were only limited. The book disproves both hypotheses and shows the broad spectrum of institutional obstacles and governmental attitudes toward East-West scientific cooperation and multiple modes of the transfer of ideas, ranging from correspondence, quotations, and books reviews, through publishing on the other side of the Iron Curtain, to participation in international conferences or joint projects.

The book contributes to a deeper collective understanding of the multiple contemporary ideological and political circumstances created by governing regimes in socialist countries and how these influenced scientific work and individual scientists' careers on the one hand and the options scientists had—and utilized—to develop their research in collaboration with their Western peers on the other. The main findings of the book are, first, that despite the common narrative of a unanimous socialist bloc under the strict supervision of the Soviets, the everyday practice of doing research and participating in the international academic community had various faces in individual socialist countries and even in individual countries in various periods. Second, many factors influenced the openness to international networks, including the distinct political and historical circumstances that conditioned how individual countries entered into the Soviet sphere of influence, the traditions of the field, and the institutional framework of science, including how scientific research support was organized. Third, for most economic historians from socialist countries who entered the international environment, the principal motivations were to develop their research, to maintain academic quality, and to win international recognition.

In its seven national chapters, the book seeks to find answers to several important questions. First, what was the nature and evolution of institutional obstacles to East-West scientific cooperation that historians on the east side of the Iron Curtain had to face? Second, what were the modes of East-West scientific cooperation and of the transfer of ideas between Eastern and Western scientists in individual socialist countries? Third, what was the role of Western historians and Eastern historians in Western exile in promoting cooperation between Eastern and Western scientists? And finally, how did the geopolitical situation of the Cold War era influence the trajectory of economic history in individual socialist countries?

The Cold War still waits for its master narrative (Reynolds et al., 2017). While initially the Cold War was viewed as a security emergency, in the 1950s it became a battle of global alliances and of political ideas. As superpower détente seemed to take hold in the 1970s, the view of the Cold War as a long-term conflict of opposed systems driven by interest clashes of the strategic security needs of great powers prevailed. From its early years, the Cold War shook the existing worldview, beliefs, and social structures on both sides of the conflict (Westad, 2010). Future socialist countries with the exception of the Soviet Union found themselves on the east front not only because of the inability of their democratic parties to oppose communists but also because of the outcomes of the Yalta conference of 1944. Despite different levels of economic development, war damages, and the necessity of postwar reconstruction, all future socialist countries were united by the geopolitical consequences of the Cold War.

While the geopolitical situation provides a context for the story and individual human fates provide examples of the complex circumstances of and complicated conditions for international cooperation, the main focus of the book is on the varieties of cooperation between Western and Eastern economic historians and the diffusion of new ideas and methodological improvements in economic history in both directions across the Iron Curtain.

Recognizing the Discipline: The Early Years of the Discipline of Economic History in Future Socialist Countries from Its Origins to World War II

There is a unique paradox in the history of economic history in Central and Eastern Europe. In the early years of the discipline in the territory that became the so-called socialist bloc after the Second World War, only one country of the seven under consideration in this book existed on the political map of Europe; today four of them no longer exist. At the end of the nineteenth century, Germany was one state. The Czech, Slovak, and Subcarpathian parts of future Czechoslovakia were parts of Austrian and Hungarian parts of Habsburg Monarchy, respectively. Poland was partitioned among Germany, the Habsburg monarchy, and the Russian empire. The Russian empire consisted of different territory than the Soviet Union after the Second World War. Romania occupied only part of its future territory, and Yugoslavia did not yet exist. Their future territories were partitioned among three empires: Habsburg, Russian, and Ottoman. This paradox helps us understand how much the evolution of the discipline of economic history was determined by historical twists and turns during the turbulent twentieth century. The countries under research in the book suffered from two world wars, repeated disintegration and integration, and never-ending changes in ideological settings and political regimes. Those upheavals affected individual human lives. The life experience of historians affected their understanding of society and the everyday scientific practice, which in turn shaped research in economic history.

The fugitive existence of socialist countries was even fortified by the long-time elusive character of the term of economic history itself. Today, economic history in Central and Eastern Europe is considered mainly a historical sub-discipline. In its early years, interest in economic history was spurred by the goals of political economy and by national policies that, targeting on catching up with the West economically, supported national "industriousness." The interest in research on economic history in Central and Eastern Europe can be traced to second half of the eighteenth century when it was prompted by the socioeconomic changes of the time, by the Enlightenment, and by the development of statistics. Responding to the needs of state administration, research was dominated by statistics and topography and addressed current affairs. The beginnings of the scientific discipline can be found in the last quarter of the nineteenth century when economic history was a component of economic theories that analyzed the development from the household economy to the national economy. At the same time, national conceptions of history were an inseparable part of the nation-building process, and history as a scholarly discipline flourished in Central and Eastern Europe. Archival sources and their utilization and preservation were considered pillars of the nation-building process.

Thus, incentives for establishing the field of economic history had two roots—political economy and Rankean historicism. The first proponents of research in economic history were statisticians and economists, often graduates from law faculties in which national economy was a part of a curriculum. The first workplaces for researching and teaching economic history were established within economic departments at faculties of law. In this regards, German historiography played a decisive role for the evolution of the discipline of economic history in Central and Eastern Europe. This was not only because the first journal with the "economic history" in its title, *Vierteljahrschrift für Social und Wirtschaftsgeshichte* [*Social and Economic History Quarterly*] was established there in 1903. All researchers in economic history in Central and Eastern Europe orbited university workplaces in Germany. A significant number of students of law from the Habsburg monarchy and the Russian empire studied at German universities that were under the influence of the German Historical School. Even though no work of representatives of the German Historical School or of Max Weber or Werner Sombart was conceptualized as economic history at that time, the future proponents of the field of economic history in Central and Eastern Europe studied mainly under supervision of economists Wilhelm Roscher in Leipzig, Werner Sombart in Berlin, or Gustav von Schmoller in Halle, Strasbourg, or Berlin and not under the proclaimed economic historian Theodor Inama-Sternegg in Vienna. When these students returned home, they kept close cooperation with German economists and translated or reviewed their books for the home readership.

However, as at that time the discipline of economic history found only modest position in the university curriculum in the Habsburg monarchy, research in economic history was often the domain of regional archivists and historians at the faculties of arts, who regarded it as a form of cultural history broadly conceived (Kolář, 2008). For instance, German historians in the Bohemian lands who were interested in economic history already in 1862 grouped themselves around the Verein für die Geschichte der Deutschen in Böhmen [Association for the History of Germans in Bohemian Lands]. When the state engaged in the area of research in the Austrian part of the monarchy, it aimed to support agricultural research in accordance with the agricultural character of the area. Thus, the *Rolnický archív* [*Peasants' Archive*] was the first journal focusing on economic history in Bohemian lands; it was first published in 1902 and renamed the *Zemědělský archív* [*Agrarian Archive*] in 1914. The first person to be awarded the *venia legendi* (position of associate professor or reader) in social and economic history was Jan Peisker (1851–1933), a historian of Czech origin, at the University of Graz (Austria) in 1901. He was influenced by Karl Lamprecht and focused on agrarian history, especially the social conditions of peasants. When the German University in Prague established its first post in economic history in 1911, it was given to Paul Sander, an economist from the University of Berlin and pupil of Gustav von Schmoller. Sander focused on the history of cities from the perspective of their economic conditions (Doležalová & Holec, 2016).

The outbreak of the economic history in Hungary can be recognized in 1878 when the term appeared in the title of the book written by Béla Weisz (1848–1945). Its establishment as a discipline was confirmed by the first journal, *Magyar Gazdaságtörténelmi Szemle* [*Hungarian Economic Review*], which was founded in 1894 and focused mainly on agricultural topics. When the journal was terminated in 1906, studies in Hungarian economic history were published in *Vierteljahrschrift fur Social und Wirtschaftsgeshichte*, which helped bring the Hungarian perspective into the fore. There was, however, no university department or post with particular focus on economic history in the Hungarian part of Habsburg monarchy. As in the Austrian part of the monarchy, research on economic history focused mainly on agricultural topics and, as such, was supported by the Hungarian Ministry of Agriculture. When Ignác Acsády (1845–1906) tried to promote an interest in economic history research, he appealed to the role of agriculture in the thousand-years-long mission of the Hungarian nation (Kövér, 2016, 243).

The Polish historiography of economic history was institutionalized in the first decades of the twentieth century after years of intensive research on economic aspects of Polish history. The institutionalization of the field is connected with the two historians, Franciszek Bujak (1875–1953) and Jan Rutkowski (1886–1949). Bujak's habilitation thesis on the medieval settlement of Małopolska is considered the first Polish work in economic history. Immediately after its defense in 1905, Bujak started to lecture on the discipline at the Jagiellonian University in Cracow (Kochanowicz, 2016). From the first years of his university career, Bujak strove to build a discipline with a unique subject (economic and historical at the same time) and methodology (positivistic, empirical, statistical). Economic history was to produce multi-layered studies of households, towns, and industrial and agricultural enterprises. The outcome of Bujak's struggle was the first school of economic history in Poland. Rutkowski developed and institutionalized the field at the University of Poznań. In his view, economic history had to be a problem-oriented discipline with a firm theoretical background. While the first point later became the background for the close cooperation of Polish economic historians with the Annales school, the second point established the brand of Polish historiography in the following decades.

The beginning of Russian economic history is also rooted in both economics and history. It is connected with the work of economist M. I. Tugan-Baranovsky (1865–1919) on business cycles in Western countries and on the development of manufacturing in Russia, and of historians P. G. Vinogradoff (1854–1925) on social and economic conditions of early medieval England and I. M. Kulisher (1878–1933) on customs policy, industry, and trade (Borodkin, 2016, 195). Research on economic history was neither professionalized nor institutionalized in southeastern Europe and in Slovakia, due to their economic and educational backwardness and lack of universities and secondary schools.

In 1918, the map of Central and Eastern Europe was redrawn. Empires disintegrated and new states were created. The Austro-Hungarian empire was replaced by Austria, Hungary, and Czechoslovakia. Great Romania, Transylvania, and some additional territories created Romania. Croatia, Slovenia, Voivodina, Bosnia-Herzegovina, Serbia, Montenegro, and Macedonia became part of Yugoslavia. Poland was made up of territories from the former German, Habsburg, and Russian empires. These geopolitical changes brought about entirely new conditions for scientific research and history writing at a national level, even though all of these new countries, with the exception of Hungary, remained multinational. History started to prevail in economic history research not only because economists were required to concentrate on urgent economic issues connected with breaking with the monarchy and building the new states. Historians also fortified their position as guides of the nation for their share in the national revival of the nineteenth century, as Robert Seton-Watson noticed immediately at that time (Seton-Watson, 1922).

Despite the pronounced intellectual isolation of German academia on the global scale due to the long-term effect of the Great War (Hesse, 2016), the German influence on the Central and Eastern European area remained strong even though the number of students at German universities was reduced as each country established new universities with departments of history and economics. From the point of view of the institutional development of economic history, the first workplaces in economic history were established in Germany: in Cologne in 1919 and in Munich in 1921 (Hesse, 2016, 99). The connections between the Russian and German tradition became even stronger when economists Vassily Leontieff (1906–1999) and Jakob Marschak (1898–1977) emigrated to Berlin. Alexander Gerschenkron (1904–1978) worked with Friedrich Hayek in Vienna, where the influence of the subjective school was growing. Austrian legal scholars developed the field of economic history under the lead of Alfred Dopsch, who untied the close ties with juridical and economic traditions. Dopsch opened the door for the dominance of history in economic history research in Austria when he established the Seminar for Economic and Cultural History at the University of Vienna in 1922. With the expulsion of 1,400 scholars in the social sciences in the second half of the 1930s, some of them found safe harbor behind the Atlantic; among them Fritz Redlich in the Center for Entrepreneurial Research at Harvard University where he cooperated with David Landes and Alfred Chandler (Hesse, 2016, 100). Russian-German economic historians Leontieff and Gerschenkron became American economists. Leontiff received the Nobel prize for his research on long-term economic—thus historical—phenomenon.

In the first years after the Russian Revolution, Russian historiography witnessed continuity in research based on rich statistical resources from the tsarist era; authors developed research on Russian, European, and global history and on the history of industry, wages, labor, enterprises, agriculture, and the new theory of Asiatic mode of production (Borodkin, 2016, 198). This theory addressed the East-West dichotomy and saw the uniqueness of eastern societies in their economic development. However, these disputations were only short-lived as the ideological orientation had been changed dramatically. While before 1917 economic historians demonstrated their non-Marxist approach by delimiting themselves from Marxist theory, after 1917 Marxist orientation became an assignment. The narrative and consequent research of the succession of five societal formations: primitive, slave, feudal, capitalist, and communist replaced discussion about the East-West dichotomy. In the 1920s, the most political figure among economic historians, Marxist M. N. Pokrovsky (1868–1932), coined the interpretation that the development of capitalism was an outcome of the competition between industrial capital and trade capital in the second half of the nineteenth century.

While in the 1920s Pokrovsky was the first to introduce purges at academic workplaces in the social sciences, later on his Marxist approach was criticized as too moderate (Borodkin, 2016, 197). The forced collectivization and industrialization of the 1930s demanded a rejection of any kind of respect for the tsarist pre-Soviet economic and social order. Socialism had to be characterized as an economic system in which all means of production were in the hands of the working class and in which technology developed faster and more effectively than in capitalism. Not only were Pokrovsky's books removed from libraries and textbooks with a new interpretation of history published. A new wave of purges took place. The purge soon revealed its negative impact on the development of the field of economic history in the Soviet Union. The leading economic historians, economists N. Kondratiev (1892–1938), famous for his theory of economic cycles, and A. V. Chayanov (1888–1937), world-known for his analysis of agriculture and cooperative movement, were sentenced to death.

In the newly established states, the discipline of economic history was pursued in two different contexts, within and outside the university setting, with the latter taking the lead. Thanks to the work of Bujak and Rutkowski, Polish economic history became a fully developed scientific discipline elaborating creatively German, French, and Marxist inspirations. Both of them established schools of economic history thanks to which by the end of the interwar period, Polish economic history was an umbrella for various subfields of history rather than a subfield of history. Chairs in economic history existed at universities in Lviv and Poznań since 1920. Economic history as an independent field had been represented at congresses of Polish historians since 1925. A specialized journal was established in 1931. In hands of Bujak, Rutkowski, and their pupils, economic history took into consideration demography and statistics and studied money, prices and labor, agriculture, trade, and industry but also the influence of exogenous factors on economic development. It produced problem-oriented studies as well as multi-aspect monographs based on theoretical concepts, comparative analysis, and statistical evidence.

In Czechoslovakia, the Committee for Agricultural History, founded in 1930, researched the most recent historical periods (from the eighteenth century onwards), conducted archival investigations, and promoted publications in the area of agrarian historiography. The importance of agrarian history was reflected in its new role as a university subject. In 1922 economist Josef Kazimour (1881–1933) started lecturing on agrarian economic history at the University of Agriculture in Prague. A decade later, historian Bedřich Mendl (1892–1940) was awarded *venia legendi* by the Faculty of Arts at Charles University (1927), where he established a Department of Medieval and Early Modern Economic and Social History in 1934 (Doležalová & Holec, 2016). The publishing platforms for economic history were established as well. *Časopis pro dějiny venkova* [*Journal for the History of the Countryside*] published widely in agrarian history. *Český časopis historický* [*Czech Historical Review*] provided historians with information about new trends in European economic history in the form of review articles. The growing interest in economic history is evident from the journal's index, which listed further subdivided categories from the 1930s (general economic and social history, handicrafts and industry, commerce, finances, and taxes).

In Hungary, the first economic historian is considered Sándor Domanovszky (1877–1955), who started to teach at the Commercial Academy in Budapest and later at the Pázmány Péter University where he redirected the interest of many students toward economic history. The first chair in economic history, that obtained Ferenc Kováts (1873–1956), was established at Budapest Polytechnic in the 1930s (Kövér, 2016, 243). Since the first years of Annales, Kováts published his reviews of Hungarian economic history historiography in the journal (Kövér, 2016, 244). Alfons Dopsch had particular influence on the development of the field in Hungary. He was elected a fellow of the Hungarian Academy of Sciences in the mid-1920s and regularly delivered lectures in Budapest on the methodology of economic history, which inspired the systemic research of Domanovszky and his pupils on the manorial economy. The outcomes of this research mirrored in fifteenth volumes of the series *Studies on the History of Hungarian Agriculture*.

Economic history was not clearly defined in Romania or in Yugoslavia. The discipline of economic history was marginalized at the edge of historiography, producing only a few studies inspired by the conviction that historical processes were determined by economic factors as well. The view was inspired by the work of German economist Werner Sombart and Russian economist A. V. Chayanov. In his major work, *Evoluția economiei românești după primul război mondial* [*The Evolution of the Romanian Economy after the World War*] published in the year of his death, economist Virgil Traian Madgearu (1887–1940) analyzed the Romanian family holding. He was attracted particularly to Chayanov's arguments about the qualitative differences between peasant husbandry and the large-scale, agricultural enterprise of a capitalist nature. In the interwar era, however, capitalism was undergoing only modest beginnings in Romania.

To summarize the early roots of the field of economic history in Central and Eastern Europe, with the exception of Germany and Poland no school of economic history developed until the period after the Second World War. Only individuals elaborated economic history issues that were closely connected with the predominantly agrarian character of their economies. Research in industrial development found its firm position only in Czechoslovakia and the Soviet Union. With the strengthening position of history in economic history, historical methodology started to dominate. The holistic view of national history that included political, social, cultural, and economic dimensions became evident in the struggle of national historiographies to write national history syntheses that had to break with pre-Great War historiography and support national state building. Fascination with national history was a common feature of Central and Eastern European historiographies as Czechs, Slovaks, Poles, Hungarians, Romanians, and Yugoslavs searched for inspiration in past greatness. In spite of the growing influence of the French Annales school (Pomian, 1978), throughout the entire interwar period the methodological inspiration came still mainly from the German sphere. In the 1930s, economic history in Germany merged with historical research, economic geography, and anthropology (Burleigh, 1988), and research expanded into the territory of Central and Eastern Europe.

A Nonuniform Voice: The Development of Economic History in Socialist Countries During the Cold War Era

As the Great War, so also the Second World War redrew the map of Central and Eastern Europe. Germany was divided into two new states, Czechoslovakia lost its Subcarpathian part to the Soviet Union, Polish state territory moved west. Germans were expelled from Central and Eastern Europe. Yugoslavia was transformed into a federation of six republics, which, with the exception of Bosnia and Herzegovina, were based on the ethnic principle. Even more dramatic were ideological changes that transformed the geopolitical situation throughout the world for the end of the Second World War brought the outbreak of the Cold War, worldwide conflict of ideas, technologies, and cultures that created a new world system, dualistic and technology-based (Westad, 2010). The Iron Curtain, both a metaphor of ideological division and a physical barrier, was erected to block the view from the East to the West and other way around.

Immediately after the communist takeovers, academia in socialist countries was reconstructed politically and socially following the pattern developed in the USSR during the interwar era. Universities were turned into "proletarian" outposts controlled by the communist parties (Górny, 2013, 39). As the reconstruction of academia took place in the years of the political fight for the gaining and consolidation of communist power, it was accompanied by a wave of purges that devastated the prewar generation of historians by the forced exclusion of those who were not willing to accept the new rules. Some of them were expelled from academia, some of them accused of anti-regime subversive activities, some of them jailed or forced to emigrate. However, the reconstruction of academia yielded divergent results in each country and was never completed in some of them. Poland differed from other socialist countries significantly due to the strengthening of its academia during the war (Connelly, 2000). In Czechoslovakia and Hungary, non-communist university teachers were forced out of their jobs or were at least blocked from advancing to the position of docent or professor or publishing books and articles. Some of them were moved from teaching to research positions within the newly established institutes of the academies of sciences or archives. In some cases, these restrictions did not apply to regions outside the capital or to less-visible institutions. In all countries, the archives of nationalized companies became harbors for excluded historians. For instance in Czechoslovakia after 1970, decentralization of historical research created a "gray zone" of historical research between official and dissident currents, members of which worked in regional and company archives. Significantly, in the peripheralized archive of Charles University, historian Jan Havránek (1928–2003) utilized his seclusion foreign students and researchers (in contradiction to official scientific policy) to realize their research in Prague.

At the same time, the attitude toward the regime and the behavior of historians during the Cold War varied. We can identify devoted Marxists who discovered Marxism already during the interwar period and believed they could apply research in favor of a better communist society. Some of them lost their devotion during the waves of political purges. A second group were the ardent Sovietizers who were willing to falsify history in favor of what they claimed was an inevitable communist future. The third, largest group oscillated between pragmatism and opportunism. They sought to prevent distortions of history, keep the profession alive, and in some cases even protect endangered colleagues. The majority of them chose to pay lip service to the regime and to continue the more or less solid research on less ideologized areas; research on the medieval era and on the feudal society was often the first choice given their long tradition in Central and Eastern Europe. Witold Kula (1916–1988), non-Marxist but leftist Polish economic historian, captured the inner dialogue of many intellectuals in an essay published in 1958. The essay was written in the form of letters of two fictional Romans living in the early years of Christianity. The fictional characters meditate on whether the new faith would bring new promises or destroy the achievement of an established civilization (Kochanowicz, 2016, 217). The loss of personal integrity, demoralization, self-censorship, and provincialization of historiography were the price the community of historians had to pay.

During the forty years of communist regimes, scientists experienced a few waves of communist purges, periods of liberalization, and ideological clashes of varying levels of intensity as the journey through socialism was not synchronized in all socialist countries. The end of Stalinism in the mid-1950s was a first crucial moment for all countries with the exception of Yugoslavia, which followed its own socialist trajectory from 1948. The USSR experienced a short period of de-Stalinization that partially broke through the self-isolation and temporarily rehabilitated pre-Soviet historiography. Czechoslovak historians breathed freely only during the short period of Prague Spring in the mid-1960s. At the same time, a set of more restrictive policies was set in motion in East Germany. In Czechoslovakia, oppression was reintroduced after 1968, strengthened in 1977 in response to *Charta 77*, and lasted until the second half of the 1980s when the Soviet Union started the era of thaw in connection with perestroika. In Romania, the Soviet-dependent period during which historians had almost no connections with the West ended in the 1950s; the liberal period followed until the end of the 1970s when Nicolae Ceausescu's regime reinforced a strict policy of ideological control. Thus, only Polish and Hungarian historians experienced a long-term era of openness to the West; in Hungary, the turning point was the revolution of 1956 and in Poland the clash between mainstream and dogmatic historians at the beginning of the 1950s that ended with the defeat of the latter. Thereafter, economic history in both countries developed in a quite open atmosphere. While in Poland the windows to the West were partly closed after the imposition of martial law at the beginning of the 1980s, the same decade of the 1980s was the most fruitful for Hungarian historiography, crowned by the organization of the World Economic History Congress in Budapest in 1982.

The above periodization informs us not only about the course of development of the communist regimes in individual countries. It informs us also about dramatically different periods that allowed scholars to research freely or nearly freely without ideological pressures in individual countries during the years of the Cold War. Ideological pressure and the influence of geopolitical circumstances were everyday aspects of life in the socialist bloc, which scientists could not avoid. Several generations of scientists experienced repeatedly internal persecution, international isolation, and multiple situations when not just their professional careers but even their ability to retain a research job depended not on their academic credentials and specialized knowledge but on subservience and clientelism. Moreover, regime surveillance had not only open but also hidden forms of oppression that could force scientists into inner exile or ban them from participating in academia.

Dramatic transformations in working conditions in science and education influenced the continuity and evolution of the field of economic history. With only a few exceptions, scientists were able to work at universities and academies of sciences if they became the communist party members, accepted state-approved curricula in education, and followed the tasks assigned by the central committees of the communist parties in their research. Under these conditions, neither economic history nor other social sciences and humanities were a free intellectual exercise on the eastern side of the Iron Curtain for two reasons. First, history had to be rewritten again. This time the new narrative had to overcome old national stories and construct the model of four consecutive modes of production that would serve the ideological needs of the new regime. Second, Marxism became the only methodological and ideological framework of history.

Addressing the role Marxism played in the evolution of the discipline of economic history, we can notice another paradox of historiography in Central and Eastern European countries: it was easier and less conflicted to be a Marxist in Western countries than in socialist countries. Moreover, Marxism was not treated in the same way in all countries and at all times during the communist era. On the one hand, universities in all socialist countries were deluged by newly established institutes of Marxism-Leninism that had to spread a new Marxist-Leninist worldview among students. On the other hand, no matter how firmly the countries were united under the banner of Marxism-Leninism in the official narrative, Marx's teaching had almost no tradition in Central and Eastern European countries. Doubtless a considerable group of historians in all countries noticed the work of Karl Marx. In accordance with public communist propaganda, Marx's teaching was understood as a method of production and distribution or as a tactic of the Second International. The official Marxism-Leninism (called Stalinism until 1956) that penetrated all levels of education oscillated within boundaries that were set by resolutions issued by communist party congresses. Thus, Marxism-Leninism was—with the exception of Yugoslavia, whose leadership was developing its own version of Marxism—the subservient application of what was the transformation of Marx's teaching to the particular conditions in revolutionary Russia developed by Lenin. In this narrative, Lenin as a successor of Marx surpassed Marx's teachings in some respects.

As a consequence, in most historical academic writings Marxism-Leninism became only a libation to a god. While the names of Marx, Engels, and Lenin (for the short period of time until 1956 also Stalin) could be found in the introductory paragraphs of every published academic book or article, Marx's historical materialism can be recognized only rarely as a value-neutral methodological tool in historical writings. The only exception was Polish historiography, in which the Marxism was considered a tool of historical analysis already in the interwar period. Significantly, a large number of those few scholars in other socialist countries who used Marxism as a methodological tool got to know Marxism in exile in the West during the Second World War or through relations with Western Marxists after the war. Consequently, the waves of purges and exclusions from academia often zoomed in on these "Western-style" Marxists.

This treatment of Marxism had broader consequences. First, with only a few exceptions, Marx's teaching mutated from a holistic critical perspective into a quasi-Marxist narrative and instrument of political power and legitimacy for the ruling powers. Second, while Polish historians mostly applied Marx's methodology but did not entirely sympathize with the communist party, Soviet, Czechoslovak, Romanian, and East German historiographies proved the very opposite: the mainstream of historians were members of the communist party but were not familiar with Marx's teaching even though a significant part of them viewed themselves as Marxists. After five years of Stalinist oppressions, only seven of 52 professors of history in Poland were members of communist party, while fourteen of the 21 original members of the Institute of History in Czechoslovakia, which was established in 1952, were communists. Another case was Yugoslavia, which left the socialist bloc and searched for a unique application of Marx's teaching that, ironically, had no special imprint on historical writings. Third, because in all socialist countries a less-oppressive attitude toward Marxist-Leninist purity can be recognized in research on older historical periods, scientists found themselves comfortable in the traditional research areas of economic history in this part of Europe. Feudalism, the manorial economy, and the early stages of capitalism thus remained the core of economic history research during the socialist era.

Vulgarization of Marxism had, however, one positive aspect for economic history. As postulates of Marxism-Leninism were taken word-for-word, so was the postulate of the primacy of the economic base over the societal superstructure. Therefore, every historian was confronted with economic history and every researched question had to be conceived from the angle of socioeconomic conditions and effects. Even though only a few historians were able to apply a holistic Marxist approach in their research and writing, works that struggled for theoretical explanations of historical phenomena, as well as those describing particular phenomena, had to pay attention to the material conditions of the era under research and include economic aspects of researched events or processes.

But paradoxically again, with the exception of Poland, economic history never occupied a central position within the organization of knowledge in the socialist bloc and was not regarded as a major historical field even though it became a compulsory part of the curriculum at economic faculties in all socialist countries. Moreover, financial priority was given to search for the revolutionary tradition of nations, workers' movements, and the history of communist parties. These subjects were taught at all universities in the socialist bloc as the history of the international workers movement and a subject called scientific communism. Jointly with Marxist-Leninist philosophy and political economy were institutionalized under the umbrella of above-mentioned institutes of Marxism-Leninism.

Yet, economic history in socialist countries achieved significant outcomes that enriched European historiography. However, we need to take into consideration the state of historiography in individual socialist countries. Even though the implementation of the Soviet model disturbed the continuity of the field in many respects, individual national traditions—either strong or weak—continued to influence the evolution of economic history during the entire communist era.

In Poland, economic history achieved international recognition already during the interwar period. The discipline had a strong position with competing workplaces and economic history schools at universities in

Cracow, Warsaw, Łódź, and Poznań, a broad spectrum of financial support, and world-known figures. Polish economic historians merged various Marxist and non-Marxist approaches and Marxism was a seriously taken methodological approach. The long-term focus on the manorial economy brought its fruits in the form of the development of original theoretical concepts. Marian Małowist (1909–1988) of the University of Warsaw explained the trade relations and economic inequalities between East and West Europe from the medieval to the modern era. Jerzy Topolski (1928–1998) of the University of Poznań invented new methodological ideas for economic history. Witold Kula of Łódź and Warsaw developed a theoretical concept of the evolution of the Polish feudal system with broad implications for general analysis of non-capitalist systems, challenging the universalist claim of one type of economic theory (Kochanowicz, 2016, 221). Thanks to its deep penetration into European and world economic history circles, the legacy of Polish historiography remains the core-periphery perspective and the concept of the economic backwardness of agricultural Eastern Europe. At the same time, Poles were the first to catch the trend of writing economic history in symbiosis with social and cultural history; Kula already in 1970 published the book on the replacement of traditional weights and measures by the metric system that combined economic, social, and cultural history with anthropology.

The focus on industrialization relates Czechoslovak economic history with East German historiography. Following the rules of denazification, the need to build a new progressive revolutionary historiography, and historians' political and ideological preferences, historians distributed themselves between the two German states (Hesse, 2016). Having no option to search for revolutionary traditions in the entire German history but forced to build up Marxist historical scholarship, East German historiography turned toward the progressive tradition of German science and the workers' movement. This focus reinforced the importance of economic history. Besides Jürgen Kuczynski's (1904–1997) professorship in economic history at Humboldt University, in 1950 a professorship in economic history held by Hans Mottek (1910–1993) was established at the newly founded University of Economics in Berlin. Within two decades, Mottek developed a workshop of a dozen of economic historians focused mainly on Central and Eastern Europe and its economic backwardness. The team produced textbooks on German economic history that were used by students at universities of West Germany as well

(Roesler, 2017). The Institute for Economic History at the Academy of Sciences established under the lead of Jürgen Kuczynski in 1956 gained even broader international recognition. Within its nine departments up to fifty researchers covered the time span from antiquity to recent periods, elaborated both industrial and agricultural history, and applied mathematical and quantitative methodology. Remarkable and innovative achievements can be found in their research on everyday life, industrialization, foreign trade, capital exports, economic crises and booms, environmental history, and the comparative economic history of East and West Germany (Fair-Schulz, 2017). From the writings it is evident that East German historiography was under the strongest influence of the dogmatic domestic nomenklatura and Soviet historiography.

In comparison with Polish and East German historiography, in other countries only small groups or isolated individuals were developing economic history as an independent field. In Czechoslovakia and Hungary, the backgrounds of the discipline that were established during the interwar period were significantly disrupted during the war. In Romania and Yugoslavia, the field was not distinguished before the war and developed only slowly afterward.

In Romania and Yugoslavia, economic history writings followed the traditional descriptive patterns, relied on archival sources, and only rarely incorporated quantitative analysis. At the core of the interest was the medieval era and the early stages of capitalism. The number of those who proclaimed themselves as economic historians was modest in both countries. Those who elaborated economic issues, such as Andrei Oțetea (1894–1977) in Romania or Nikola Vučo (1902–1993) or Mijo Mirković (1898–1963) in Yugoslavia, did so as a part of broader historical writing; historians dominated economic history research. The development of economic history in Yugoslavia was retarded by its captivity to national frameworks at the level of individual republics. In Croatia and Serbia, republics with longer historiographical traditions, the subject of economic history was introduced in the 1960s at the faculties of history at the universities in Zagreb and Belgrade. The division of research between economists and historians retarded the development of Romanian historiography. In 1970, the Committee of Economic History and History of Economic Thought were established as the framework in which Romanian historians and economists were expected to cooperate. The aim was not accomplished.

The same can be said about Czechoslovakia, where economic historians teaching economic history at universities of economics were underestimated and even overlooked by historians from the faculties of arts and institutes of history of the Academy of Sciences, for they had no historical training. But in fact, in Czechoslovakia the first specialized workplaces in economic history were founded at universities of economics in Prague and Bratislava in the 1950s. At institutes of the Academy of Sciences in Prague, Bratislava, and Opava only small groups of historians elaborated economic history. Besides them, only isolated individuals—historians by education—operated at faculties of arts and in archives of nationalized companies. While economic historians from universities of economics concentrated on writing descriptive textbooks, inspiring discussions about prices emerged within the Working Group on the History of Prices and Wages (established in 1958), which focused on tracing long-term price trends for various commodities, their contextualization, and the extent to which markets operated. While scholars involved in this debate professed inspiration from Soviet, Hungarian, and Polish historiography, individual studies show clear signs of the influence of the Annales school and its research on consumption and the standard of living. Industrialization was another subject of a dispute in the Czechoslovak environment that was opened by the Slovak economic historian Anton Špiesz (1930–1993), who disagreed with Czech economic historian Jaroslav Purš (1922–1997). Purš soon became the most important figure of Czechoslovak economic history thanks to his firm position in the communist nomenklatura. Outside of the mainstream but provoking strong criticism from the nomenklatura and positive feedback from abroad, Milan Myška (1933–2016) elaborated the concept of proto-industrialization and Alice Teichova the role of foreign capital in interwar Czechoslovakia.

After the war, the Hungarian school of economic history was established at the Department of Economic History at the Karl Marx University of Economics by Pál Zsigmond Pach (1919–2001), a former student of the interwar promotor of economic history, Domanovszky. Thanks to the firm interconnection of curricula in economics and history, a new generation of Hungarian Marxist economic historians was educated in economics and history at the same time; among them the future prominent figures of Hungarian historiography of economic history, Iván Berend (1930) and György Ránki (1930–1988). Research focused mainly on the nineteenth and twentieth centuries, the evolution of capitalism in Hungary, and the economic performance of the Habsburg empire motivated by Central European sentiment. By the early 1980s, Marxism-Leninism no longer figured in books on history, not even as a brief invocation of Marx, Engels, and Lenin. Instead, historians embraced the Annales school, American new economic history, the Bielefeld school, and Foucauldian inspirations (Kövér, 2016). As the straitjacket of Marxism-Leninist loosened, Hungarian historiography maintained its cooperation with foreign workplaces. These achievements were crowned by Ránki's chair in Hungarian Studies at Indiana University in Bloomington and his commuting between the United States and Hungary throughout the 1980s. This would not have been possible in another socialist country.

In the Soviet Union, various trajectories of continuity were exhibited by Russian and other historiographies. During the Stalinist purges and war, Ukraine lost three-quarters of the economic historians who were educated at the end of the 1920s. The Ukrainian case demonstrates the dramatic consequences integration into the Soviet social, economic, and political core could have for historiography; the restoration of historiography took longer in Ukraine than at the central level represented by Russian historiography. Already in 1944 the Presidium of the Academy of Sciences drew up a list of the principal problems of historiography requiring immediate attention; among them were the formation of the all-Russian market of the feudal society, its development, the degree to which commodity-money relations had spread in the feudal economy, and the transition to new capitalistic relations (The Principal Problems, 1947). The first workplace in economic history was established three years later as the Department of History of the National Economy and Economic Thought in the Faculty of Economics at Moscow State University chaired by A. D. Udaltsov (1883–1958). As the department was established during the campaign against objectivism and cosmopolitanism in social sciences, it was not possible to accept any foreign or pre-Soviet historiographical traditions and was unacceptable not to quote writings by Stalin and classics of Marxism-Leninism. The main mission of economic history was to demonstrate the achievements of the Soviet and socialist economy and its supremacy over the capitalist system. The study of pre-revolutionary Russia was allowed only in the search for evidence of the inevitability of the Russian Revolution. Works that elaborated the period of the New Economic Policy, collectivization, or industrialization of the 1920s and 1930s had an apologetic character.

During the thaw after the Twentieth Congress of the Communist Party of 1956, the work of excluded economic historians (including Pokrovsky's) was partially rehabilitated. However, only after the Leningrad congress in economic history in 1970, where more than of half of thousand delegates were Soviet scientists, did economic history obtain the attention of the Communist Party as a discipline on the edge of history and economics that could help understand better the social and economic aspects of history. As a consequence, in the 1970s and 1980s economic history started to flourish, elaborating new topics of research in agricultural history, financial and banking history, history of the labor market, and history of trade and using quantitative methods. General works on Russian economic development with special focus on the late nineteenth and twentieth centuries were published. For a short time discussion about the East-West dichotomy found its way into the mainstream, utilizing the anti-colonial movement in which historians from other socialist countries participated as well (Borodkin, 2016, 201). At this moment, economic history was mostly in the hands of historians; the official narrative was that the study of history distracted economists from the search for solutions to current economic issues (ibid., 202).

Different traditions of economic history in individual countries were also visible in the spectrum of specialized journals. While in Poland or the Soviet Union, journals established during the interwar period continued to be published after 1945, in other socialist countries—and at the republican level in the USSR—new specialized journals had to be established as the entire institutional background of the field. These new journals, however, did not achieve international recognition. This was partially because they were published in national languages, and only papers to be presented and distributed at World Economic History Congresses were published in English-language issues.

To summarize the achievements of the historiography of economic history in socialist countries, first, economic history was elaborated as a part of the broad social, political, and cultural history. This can be demonstrated by the long-term struggle—not always successful—of national historiographies to produce multi-layered national history volumes (History of Poland, Czechoslovak History, History of the Peoples of Yugoslavia). Second, the discipline of economic history was a sub-discipline of history rather than of economics and was defined in thematic rather than methodological terms. Third, the majority of historians defined themselves as historians with research (also) in social and economic aspects of the researched topic. It is, therefore, impossible to count the number of economic historians in individual countries or to enumerate the economic-historical studies that were published in individual countries of the region for they were scattered across a number of sometimes surprising institutions having nothing obvious to do with economic history. They were, however, united by the publicly touted conviction about the Karl Marx's holistic approach to history. Even though the official ideology of Marxism-Leninism penetrated academia and compromises were necessary in historical writings, in economic history Marxism-Leninism usually was but a few quotations from classics in the first paragraphs and proper anti-capitalist phrasing in following pages that, occasionally, could provide fresh and original intellectual insights. The vast majority of writings were descriptive books that provided material based on archival research and quantitative data in the form of simple statistics. At the same time, quantitative tools of research were limited to statistics, only a limited number of studies extended beyond a short period of time, and even fewer analyses sought to explain the process of historical change.

In fact, the Marxist holistic approach was creatively elaborated only by a few schools of economic history in Poland and East Germany, and by a few historians in Hungary, Czechoslovakia, and in the Soviet Union. Thanks to these scholars, historiography introduced innovative topics that became fashionable in world and global historiography decades later—Polish historians' core/periphery theory, German comparative/global economic history, or Czechoslovak Miroslav Hroch's theory of national movements. When assessing the relevance, or even the lasting value of historiography of economic history in socialist countries, it is thus necessary to emphasize the distinctions among the inspiring global aspirations of Poles, the visible Central Eastern European ambitions of Hungarians, and the mostly national milieu of other national historiographies.

Heading West: Factors and Modes of the International Cooperation of Socialist Economic Historians During the Cold War Era

The tradition of economic history in individual countries, besides geopolitical and ideological circumstances and an institutional framework based on scientific policies, played a key role in developing international cooperation of socialist countries. Scientific policies depended not only on declarations of the communist parties at their congresses but also on the structure of the political regime. Countries that were organized federatively—the Soviet Union, Yugoslavia, and Czechoslovakia—distributed the responsibility for scientific research differently among the particular governmental levels. These countries had to deal with the existence of multiple historiographies as well. Yugoslavia proceeded from centralization to decentralization of the scientific research agenda at the republic level. Czechoslovakia, even after including federalization in its constitution in 1968, pretended until 1989 to consist of one unified nation with one institutionally framed historiography, despite distinctions between Czech and Slovak historiographies. Strictly defined "national" policy in the Soviet Union created ambiguous relations between Russian and other republics' academic institutions.

Thanks to the firm institutionalization of the discipline, ideas developed by Polish scholars flowed internationally. In comparison with Polish historiography, in other countries only isolated individuals elaborated research internationally in cooperation with Western colleagues, and only rarely did they achieve international recognition. In Hungary, a few strong figures were able to achieve influential positions in international structures, thanks to which the second World Economic History Congress in the socialist bloc took place in Budapest. By contrast, Czechoslovaks, despite the significant role of Arnošt Klíma (1916–2000) in international networking from the 1960s, were not able to dynamize the field toward broader international cooperation. The same can be said about East German historians even though they developed two outstanding research institutions and their Marxist narrative of colonial studies and their analysis of multinationals became influential not only in the socialist bloc.

Another paradox must be mentioned here: despite the existence of numerous bilateral and multilateral agreements, cooperation within the communist bloc in social sciences was not fully open or free of conflict. The so-called proletarian internationalism, an official term that was used to describe the nature of cooperation within the socialist bloc, was only a propagandistic concept. Significantly ambiguous were relations with both German states, influenced not only by the geopolitical setting. In Poland and Czechoslovakia, the "German question" became crucial for historiography as Polish society struggled to fix its western border and Czechoslovak society needed a narrative to justify the expulsion of Germans. In both cases, historiography was critical toward Germans as age-old enemies of the Slavic nations in Central Europe (Górny, 2013, 34–35). Painful memories of Nazi occupation were still present on the individual level, which created obstacles for cooperation with Germans regardless of which state. The chair of the Czechoslovak-GDR Committee of Historians František Graus (1921–1989), who spent the war years in the ghetto in Terezín, had a very reserved attitude toward East Germans, while Jaroslav Pátek (1934–2003), whose uncle was tortured to death in a concentration camp, cooperated with no biases from the end of the 1950s with historians from the GDR on Alice Teichova's team (Doležalová, 2021).

In 1953 in Budapest, the tradition of joint meetings of the highest representatives of socialist historiographies was established. These meetings took place in the framework of congresses of national historical societies. No matter which country organized the meeting, the Soviets played the main role. The aim of the first meeting in Budapest was to break the isolation of socialist historiographies and challenge the dominion of its "bourgeois" counterpart. More importantly, in accord with Stalin's thesis from his *Economic Issues of Socialism in the USSR* of 1951, the meetings had to support the progress of socialist historiographies through joint plans of research. In the first plan, scheduled in Budapest, the anniversaries of the Russian Revolution of 1905, the liberation of Central and Eastern European countries from the Nazis in 1945, and the Russian Revolution of 1917 were listed as the main research agenda for all countries for the following years. The Czechoslovak delegation suggested to elaborate also the beginning of both feudalism and capitalism in the Bohemian Lands, the Hussite movement, the working-class movement, and the fight against the nationalism and chauvinism. In Budapest, countries also signed bilateral contracts on mutual scientific cooperation that addressed organizing conferences, writing reports on findings of other socialist historiographies in the main historical journals in each country, and inviting scholars from other socialist countries for archive visits (Macek, 1953).

The main agenda of these meetings, however, remained joint preparation for world congresses. A turning point was the year 1970 when both historical sciences and economic history congresses were held in the USSR. As the congresses took place in the week of the second anniversary of the occupation of Czechoslovakia, the participation of historians from socialist countries was carefully prepared and managed: topics were discussed in detail, historians were instructed to minimize their contributions to discussion and to reply only briefly to questions during the panels, and to converse about political issues was strictly forbidden. To achieve the conflict-free course of the congresses, historians from socialist countries had to have numerical superiority (Jiroušek, 2006, 77). The tradition of holding pre-congress meetings among socialist historians continued until 1989 and in less-open countries (the GDR, Czechoslovakia) influenced the composition of delegations for the highest international events as well.

When eventually economic historians in socialist countries could explore the way to the West, they had five key modes. They could read foreign books and journals, correspond with foreign colleagues, utilize relationships with exiled colleagues, meet foreign scholars arriving to their country, and publish their books with foreign publishers abroad or travel abroad for conferences, joint projects, or visiting fellowships.

However, traveling to the West was a privilege that could not be given to everyone. The governments of the GDR, Czechoslovakia, and the USSR shared the fear that historians would criticize the regime abroad, even inadvertently, and therefore governments in these countries introduced strict measures for the selection of representatives to travel to congresses and for visits abroad. Czechoslovak scientists could not travel completely freely even during the period of Prague Spring in the second half of the 1960s. Despite the more liberal nature of the period from the late 1950s to the early 1970s, travel abroad was a controlled privilege also in Romania. In the Soviet Union, Russian historians were given preferential treatment to the detriment of representatives of republican historiographies. While the regime in Yugoslavia did not create obstacles to travel for research or study abroad, and in Hungary and Poland scholars and students could take an advantage of foreign scholarships from the mid-1950s, for Czechoslovak, East German, or Soviet historians these doors were opened only for short periods and just for PhD candidates and senior scholars.

Students from the West traveled to Hungary, Poland, and Yugoslavia but rarely to Czechoslovakia, Romania, or East Germany and even then mainly for archival research. The same can be said about the intensity and nature of visits of Western scholars to individual socialist countries. In Czechoslovakia, Romania, East Germany, or the Soviet Union, visiting foreign scholars were monitored by the state police (Verdery, 2018) and could meet only a few carefully selected scientists, while in Hungary, Poland, or Yugoslavia they could meet representatives of broader academia and also students. In all countries of socialist bloc, the Annales school and Fernand Braudel in particular had a special position. Braudel visited all of the socialist countries and searched for new contacts and cooperation.

Also the openness of historiography toward foreign inspiration had a dramatically different face in individual countries. In Poland, Hungary, or even East Germany, the Annales School, historical sociology, or other new methodological developments in the field could inspire historians, while in Czechoslovakia or the Soviet Union these innovations were labeled bourgeois anachronisms and often officially prohibited. Even so, using the narrative of criticism, specialized journals published by academic authorities could provide information about Western historiographical trends. Every country also introduced different rules for purchasing foreign literature and publishing with foreign publishers. Foreign journals and books were available in libraries in some but not all countries. Surprisingly, journals were available broadly in one of the most closed countries, Czechoslovakia, and significant academic books in the field were available in Russian translation in Moscow.

As the nature of dictatorship and the economic development of individual countries forced scientists into exile in a few waves, the exiled historians could also play a role in opening channels for international cooperation. However, the waves of exile and their motivations varied across individual countries as did the cooperation with exiled historians. While scientists were leaving from Yugoslavia for economic reasons in search of better opportunities in their careers, emigration from Hungary was insignificant with the exception of the year 1956, and emigration from Czechoslovakia was politically motivated. Thus, while emigration from Yugoslavia increased with a worsening of the economic situation in the 1960s and 1980s, emigration from Czechoslovakia followed political upheavals: the communist coup d'état in 1948 forced into exile historians who did not agree with the new communist regime and the Soviet invasion of 1968 forced into exile even some of those who had previously supported the communist regime. According to the findings of the Czech Academy of Sciences, approximately 2,000 scientists emigrated from Czechoslovakia in the twentieth century, which was 1–1.5% of all scientists (Štrbáňová & Kotlán, 2011).

No doubt, the exile community had the potential to play a crucial role in building East-West scientific relations through maintaining personal and scientific relationships with their former colleagues back home.

The real influence of exile on historiography at home, however, was different in individual countries. There were problematic relations with the large diaspora in the Ukrainian case and minimal relations due to the fears of punishment of home historians back in Czechoslovakia. There were friendly interactions between exile and home historiography in the Hungarian case and critical disputes in the Polish case. For Yugoslav historiography immigration was not an issue, for East German historians émigrés were enemies. While scientists who left Yugoslavia, Hungary, or Poland continued to cooperate with their former colleagues and even supported them, Czechoslovak, East German, or Soviet historians were not allowed even to speak with exile historians when they met them abroad, mutual correspondence was carefully watched by secret police, and cooperation tended toward zero.

At the same time, many charitable institutions supported scientists in exile. National diasporas established institutions in Western Europe and in the United States to support scientists in exile. For instance, the Society for Science and Arts in America was established in New York in 1957 to help the young generation of Czechoslovak exiles in scientific careers in the U.S. Historians created a significant part of the membership, and historical issues dominated the programs of congresses the society organized every two years; these congresses became soon a meeting platform for Czechoslovak scientists in exile (Doležalová, 2018). The Jan Hus Educational Foundation was established in the UK in 1980 to support Czechoslovak scholars and students expelled from the universities and the Czechoslovak Academy of Sciences during the purges after the Soviet invasion of 1968. Stronger political pressure was experienced by those who supported Czechoslovak scholars in Czechoslovakia in the years of normalization (Day, 1999). Many visiting students, scholars, and even tourists brought foreign books to socialist countries often thanks to the support of foreign foundation.

At the beginning of the Cold War, however, it was not an easy task for the West to find a proper tool to penetrate the Central and Eastern European space convinced of the high level of national cultures and already deluged by Soviet cultural nostalgia. Moreover, U.S. propaganda doubted the usefulness of implementing cultural policy in postwar Europe. Only the establishment of the United States Information Agency in the 1950s started to change the situation (Gienow-Hecht, 2010, 406–407). From that time, many charitable and government-led institutions supported cultural exchange between the East and West. The Fulbright program, the Rockefeller and Ford foundations but also the British Council and Humboldt program provided financial support for Central and Eastern European scientists. For example, the Rockefeller Foundation financed subscriptions to forty American scientific journals to be distributed to Yugoslav universities and institutes. Thanks to this support, the American influence in the Yugoslav science grew significantly (Barghoorn, 1967).

At the same time, the security services of Western countries carefully observed the activities of Western scientists and institutions that supported East European emigrants and academics in socialist countries; historians in exile were hired to cooperate with spy agencies. For Western scientists and scientific associations, it was not easy to find the balance between a natural support for their peers in need behind the Iron Curtain and the fear of disrupting the fragile status quo in international cooperation (not only scientific) in which the Soviet Union was the most important player on the eastern side of the Iron Curtain. The approval of Moscow and Leningrad as the sites of the congresses of historical sciences and economic history in 1970 despite the occupation of Czechoslovakia demonstrated the determination of the associations not to connect science with politics; only the British, American, and Netherland committees raised objections.

The Plan of the Book

Despite the different characteristics of openness, national traditions, the maturity of the field, the state of Marxism, or the overall place of economic history in the structure of organized research in individual socialist countries, there were striking links among their economic history historiographies during the Cold War era. Life stories and the scientific careers of economic historians intertwined with geopolitical ruptures. In their endeavors to gain professional prestige, economic historians had to fight international isolation. Their achievements are inseparable from the cultural and academic conditions under which they had to work. The seven following chapters are organized according to countries' openness to international cooperation. Even though all countries experienced times of openness toward the West, the Soviet Union, East Germany, and Czechoslovakia were significantly less open than Poland and Hungary; Romania and Yugoslavia were mostly open to the world, and historians did not struggle with obstacles for a significant part of the socialist period, but the field of economic history was in its infancy in these countries.

The second chapter focuses on the Soviet Union as a country that in significant measure influenced political, economic, and social processes in the entire socialist bloc, including scientific research. Its special focus on Ukrainian historiography reveals that the relations between national and Soviet levels of governing science were ambiguous. The Soviet chapter analyzes the mutual contacts and cooperation between Soviet and Western economic history using as an example two Ukrainian schools of historical writing—Soviet republican and diaspora—established on both sides of the Iron Curtain. The East German and Czechoslovak historiographies (Chapters 3 and 4) shared a geographic position at the border with the West, a strong dependence on Soviet patterns, and only limited opportunities to cooperate with Western colleagues, to obtain scholarships, to travel to the West for conferences, or to publish books abroad. The East German chapter emphasizes the differences between a more creative, pluralistic, and innovative Marxism embodied by an older generation of scholars who were largely educated before the GDR was formed as well as the students that they mentored and trained within the GDR, on the one hand, and the far more formulaic, sterile, and narrow Marxism-Leninism created and propagated by the communist party apparatus, on the other. The Czechoslovak chapter explores the institutional framework for international cooperation as well as individual examples of cooperation through careful observation of the trajectories of development of historiography in other countries, direct cooperation with foreign scholars, or participation in conferences abroad. The Czechoslovak case studies focus on Alice Teichova, Miroslav Hroch, and discussion of the Industrial Revolution.

By contrast, historians in Hungary and Poland (Chapters 5 and 6) did not face obstacles to international cooperation from the mid-1950s; in Poland until the early 1980s, in Hungary for the rest of the era. Besides other aspects of mediating between East and West, the Hungarian chapter focuses particularly on the evolution and realization of the initiative for the World Economic History Congress in Budapest in 1982. The Polish chapter aims to identify and explain the impact of Polish economic historiography on the post-1945 international academic debate concerning the origins of capitalism in early modern Europe.

Romania and Yugoslavia, which conclude the book (Chapters 7 and 8), shared the fate of countries whose leadership opposed the Soviets; Yugoslavia even stopped its cooperation with socialist countries at the end of 1948. As a consequence, scientific communities had to bear the costs of reduced financial and networking support from the Soviet Union and the socialist bloc and had to search for new academic opportunities. The Romanian chapter highlights the contributions of Andrei Oțetea and Gheorghe Zane to economic history research in Romania, the methodological limitations of their approach, and the impediments to building up an integrated and autonomous field of economic history. The Yugoslav chapter focuses on circumstances that constituted obstacles to the methodological and conceptual modernization of economic historiography, despite opportunities for travel and liberalized cultural transfer.

The Epilogue considers whether some aspects of the socialist historiography legacy are still visible in individual national historiographies that were part of the socialist bloc.

References

Barghoorn, F. (1967). Cultural Exchange Between the Communist Countries and the United States. *The Annals of the American Academy of Political and Social Science, 372*, 113–123.

Berg, M. (2015). East-West Dialogues: Economic Historians, the Cold War, and Détente. *Journal of Modern History, 87*(1), 36–71.

Borodkin, L. (2016). Economic History from the Russian Empire to the Russian Federation. In P. Hudson & F. Boldizzoni (Eds.), *Routledge Handbook of Global Economic History* (pp. 195–213). Routledge.

Burleigh, M. (1988). *Germany Turns Eastward: A Study of 'Ostforschung' in the Third Reich*. Cambridge University Press.

Connelly, J. (2000). *Captive University: The Sovietization of East Germany, Czech, and Polish Higher Education, 1945–1956*. University of North Carolina Press.

Day, B. (1999). *The Velvet Philosophers*. Claridge Press.

Doležalová, A. (2018). *A History of Czech Economic Though*. Routledge.

Doležalová, A. (2021). *Mikuláš Teich (1918–2018): Moje století. Intelektuální biografie v dialogu*. Galén.

Doležalová, A., & Holec, R. (2016). Continuity and Discontinuity in the Czech and Slovak Historiographies. In P. Hudson & F. Boldizzoni (Eds.), *Routledge Handbook of Global Economic History* (pp. 230–241). Routledge.

Fair-Schulz, A. (2017). The Dissolution of the Institute for Economic History at the Academy of Sciences. In A. Fair-Schulz & M. Kessler (Eds.), *East German Historians Since Reunification* (pp. 155–164). SUNY Press.

Gienow-Hecht, J. C. E. (2010). Culture and the Cold War in Europe. In *The Cambridge History of the Cold War. Vol. 1. Origins* (pp. 398–422). Cambridge University Press.

Gienow-Hecht, J. C. E., & Donfried, M. (Eds.). (2010). *Searching for a Cultural Diplomacy*. Berghahn Books.

Górny, M. (2013). *The Nation Should Come First: Marxism and Historiography in East Central Europe*. Peter Lang.

Hesse, J. O. (2016). The Legacy of German Economic History. In P. Hudson & F. Boldizzoni (Eds.), *Routledge Handbook of Global Economic History* (pp. 97–113). Routledge.

Hudson, P., & Boldizzoni, F. (Eds.). (2016). *Routledge Handbook of Global Economic History*. Routledge.

Jiroušek, B. (2006). *Česká a československá účast na mezinárodních kongresech historických věd*. Jihočeská univerzita.

Kochanowicz, J. (2016). A Periphery at the Centre of Attention. In P. Hudson & F. Boldizzoni (Eds.), *Routledge Handbook of Global Economic History* (pp. 214–229). Routledge.

Kolář, P. (2008). *Geschichtswissenschaft in Zentraleuropa. Die Universitäten Prag, Wien und Berlin um 1900*. Akademische Verlagsanstalt.

Kövér, G. (2016). Crossroads and Turns in Hungarian Economic History. In P. Hudson & F. Boldizzoni (Eds.), *Routledge Handbook of Global Economic History* (pp. 242–257). Routledge.

Macek, J. (1953). Mezinárodní kongres historiků v Budapešti. *Československý časopis historický, 1*(3), 410–416.

Pomian, K. (1978). Impact of the Annales School in Eastern Europe. *Review (Fernand Braudel Center), 1*(3/4) (Winter–Spring), 101–121. Research Foundation of State University of New York for and on behalf of the Fernand Braudel Center.

Reynolds, D., et al. (2017). Probing the Cold War Narrative Since 1945: The Case of Western Europe. In K. H. Jarausch (Ed.), *The Cold War: Historiography, Memory, Representation* (pp. 67–82). Walter de Gruyter GmbH.

Roesler, J. (2017). The Dissolution of East German Economic History at the Economic University in Berlin-Karlshorst: A Typical Anschluss Procedure. In A. Fair-Schulz & M. Kessler (Eds.), *East German Historians Since Reunification. A Discipline Transformed* (pp. 147–154). SUNY Press.

Seton-Watson, R. W. (1922). *The Historian as a Political Force in Central Europe*. School of Slavonic Studies in the University of London, King's College.

Štrbáňová, S., & Kostlán, A. (Eds.). (2011). *Sto českých vědců v exilu*. Academia.

The Principal Problems of History in the U.S.S.R. (1947, January–February). *Synthese, 5*(9/10), 425–429.

van Dongen, J. (Ed.). (2015). *Cold War Science and the Transatlantic Circulation of Knowledge*. Brill.

Verdery, K. (2018). *My Life as a Spy: Investigations in a Secret Police File*. Duke University Press.

Westad, O. A. (2010). The Cold War and the International History of the Twentieth Century. In *The Cambridge History of the Cold War. Vol. 1. Origins* (pp. 1–19). Cambridge University Press.

Zadencka, M., Plakans, A., & Lawaty, A. (Eds.). (2015). *East and Central European History Writing in Exile 1939–1989*. Rodopi.

CHAPTER 2

The USSR: The Ambiguous Case of National Historiography

Volodymyr V. Kravchenko

The chapter devoted to the Soviet Union would have been easier to write some thirty ago rather than today. Until 1991 the USSR was viewed by the majority of analysts in the West as a monolithic state, which many identified with "Russia." Currently, Soviet historical legacy has been "nationalized" by the successor states to the Soviet Union from the perspective of their respective nation-state building, among which Russia and Ukraine remain the most active. However, the Soviet Union as well as its constituent republics, including the Russian Federation, were socio-cultural phenomena too complex to be described within the framework of a modern national paradigm.

The Soviet experiment was a peculiar combination of Marxist-Leninist utopia and Russian imperial legacy, which also incorporated certain elements of the local national traditions. It was a state-in-progress, which followed the zigzagging path continuously re-charted by the political

V. V. Kravchenko (✉)
University of Alberta, Edmonton, AB, Canada
e-mail: vkravche@ualberta.ca

A. Dolezalova and C. Albrecht (eds.), *Behind the Iron Curtain*, Palgrave Studies in Economic History,
https://doi.org/10.1007/978-3-031-31578-7_2

leadership. When Stalin's regime started its "Great Retreat" policy, it did not lead to the dissolution of the Soviet republics, established during the 1920s on the national basis. However, the republics were built in the highly centralized and ideologically driven imperial-like hierarchy. Gradually, the "red line" between "Sovietness" and imperial "Russianness" became so thin that sometimes it was difficult to recognize it at all.

The institutional model of Soviet historical writing was established during the era of Stalinism. It was part of the highly centralized and ideologically driven mechanism of Soviet academic life with its "sharp distinction between research and teaching, and an interpenetration of political and professional cultures" (David-Fox & Péteri, 2000, 5). Historical writing was considered an important ideological instrument of the Soviet project. In the hierarchy of the humanities and social sciences, historiography stood next to the "sacral" disciplines of the history of the Communist Party of the Soviet Union (hereafter the Party) and Marxist-Leninist philosophy and political economy. It is no wonder that historiography was closely controlled and monitored by the professional academic and ideological structures. In addition, a strict division of labor was established between the republican and central academic institutions.

The development of Soviet historical writing underwent several dramatic changes during the Cold War. During late Stalinism, it was overwhelmed by the growing wave of Russian imperial nationalism, self-isolationism, and xenophobia. The period between the middle of the 1950s and the end of the 1960s was marked by the process of partial de-Stalinization and the last attempts to revive early Soviet utopianism. This policy was accompanied by the end of isolationism, the selective rehabilitation of pre-Soviet historical legacies, and the search for a more effective balance of power between the center and peripheries. Lastly, with the end of communist experimentation in the 1960s, the country resumed its gradual departure from the utopian doctrines and practices and toward a more pragmatic, Russian imperial type of internal and foreign policy. The era of Gorbachev's perestroika belongs to yet another historical cycle.

Economic history in the Soviet Union initially was not differentiated from general history as a separate discipline. According to Peter Gatrell and Robert Lewis, in the Soviet Union, economic history "...lacked the corporate identity which it has in Britain and the United States" and "had no separate identity within the network of historical research institutes under the principal Soviet research organization, the Academy of Sciences of the USSR" (Gatrell & Lewis, 1992, 743). There was neither a Soviet equivalent of the *Journal of Economic History* nor a professional organization for economic historians. The explanation for such a state of affairs can be found in the dominant ideology. Why highlight a separate economic direction if historical science itself had been constructed on a methodological basis of the priority of socioeconomic processes over all others? What else could a Marxist history be in a socialist country, if not an economic one?

The study of the history of the Soviet/Russian economy as well as that of the national republics was conducted not only by historians, but also by economists. The Institute of Economic Studies of the Academy of Sciences of the USSR (hereafter the Soviet Academy), with its separate units devoted to economic history and the economic thought of Russia/USSR, was the leading center of that field of research. Thanks to the initiative of leading Soviet economists, the economic history of the USSR became a growing industry in the postwar years (Borodkin, 2016, 195–213; Bovykin, 1996, 7–27). At the source of this initiative stood such renowned specialists as M. V. Ptukha (1884–1961), P. I. Lyaschenko (1875–1955), S. G. Strumilin (1877–1974), and P. A. Khromov (1907–1987). The first two belonged to the vaguely defined group of scholars whose identity and profile were shaped by the imperial rather than a national (either Russian or other nation) context.

The concept of cooperation between Western and East European scholars during the Cold War can be applied to Soviet academia with great reserve for several reasons. First, the tendency toward isolationism prevailed in the policy of Soviet leadership until the end of the Soviet Union. Second, international cooperation was monopolized by the Russian metropolitan scholars who were at the top of the academic pyramid but did not represent the entire community of Soviet scholars. Third, even those privileged specialists were squeezed into a Procrustean bed of the rigid ideology and driven by current political demands rather than by scholarly requirements. Soviet historians, both in the center and in the provinces of the Soviet state, were obliged to produce canonical, normative texts designed to corroborate the Soviet historical mega narrative with carefully selected and arranged facts. That was perhaps the main difference between Soviet and Western communities of scholars.

In what follows, I will approach the issue of Soviet economic historians' encounters with Western academia from a Ukrainian perspective. I believe that the Soviet Ukrainian experience, no matter how untypical it may be, can add some important nuances to the history of intellectual communication between the two opposing blocs during the Cold War. I assume that due to Soviet Ukraine's double insulation within the Soviet/Russian imperial shell, the Ukrainian part of this story may be better understood in the context of "two Ukraines" divided by the "Iron Curtain." They were represented, accordingly, by Soviet Ukraine and the post-World War II Ukrainian diaspora. First, I will describe briefly the institutional framework of Soviet and Ukrainian historical studies in which one can search for examples of a dialogue between the main actors. Second, I will bring to the fore several examples of how Soviet and non-Soviet Ukrainian scholars attempted to establish mutual contacts during a brief period of Soviet Thaw. The chapter will end with an analysis of those topics that contained the potential for an exchange of ideas. I would be satisfied if the reader found an explanation for why the cooperation of Ukrainian Soviet scholars with their Western counterparts failed rather than vice versa.

Soviet Institutional Framework and Ukrainian Studies

Soviet/Russian historical writing demonstrated its ability to survive in the volatile and hardly predictable political environment (*Sovetskaia istoriografia*, 1996). According to Denis Kozlov, the professional culture and academic traditions of Soviet-Russian historians, which were inherited from the pre-1917 era, effectively survived Stalinism (Kozlov, 2011, 387–388). Historical science in the both Soviet/Russian capitals—Moscow and Leningrad—showed a clear historical continuity. This can be explained partially by the fact that the best Soviet specialists in various fields of the humanities and social sciences were given an opportunity to transfer from the republican to metropolitan centers.

Ukrainian Soviet historiography reflected the ambiguous status of the republic in the Soviet Union (Udod & Yas', 2020; Yaremchuk, 2009). It could be interpreted in both national and territorial contexts. On the one hand, the republic possessed the attributes of national statehood including official language, symbols, institutions, and even official representations abroad. Soviet Ukraine was recognized officially as a founding member of U.N. even though the republic authorities were deprived of any right to execute direct diplomatic relations with other countries. On the other hand, postwar Ukraine was increasingly integrated into the Soviet social-economic and historical core and was succumbing to intense cultural Russification. In spite of its deep integration in the Soviet/Russian economic and cultural space, or perhaps because of it, Ukraine's real status in the Soviet Union was unofficially recognized as "the second among equals," next to the Soviet-Russian Federation. The price for that status was high: it required rejecting modern "Ukrainian" national identity in favor of a pre-modern "Little-Russian" one. The former was incompatible with the imperial Soviet-Russian discourse while the latter fit the framework of imperial "Russianness." After a brief period of experimentation of the 1920s, Soviet Ukraine became a reincarnation of imperial "Little Russia" behind the façade of "Ukrainian" institutions and official rhetoric.

The main place in the structure of Soviet historiography was occupied by the Institute of History of the Soviet Academy. It embraced both the history of the Soviet Union and world history. The institute was structured into divisions and other units according to the theory of Marxist historical socioeconomic formations, each specializing on a historical epoch from ancient times to the contemporary period. The first place in the hierarchy clearly belonged to the twentieth century, with a particular emphasis on the Soviet epoch, starting in 1917.

The post-Stalin Soviet Union saw a dramatic increase in the institutional infrastructure of the discipline of history. The Institute of History grew from a mere 150 historians in 1946–47 to some 400 by the time it was divided in 1968. They were the cream of a profession that totaled 17,000 in 1961 (Markwick, 2001, 64). There were also various academic commissions and scholarly councils (nauchnye sovety) devoted to topics established in accordance with ideological and current political priorities: the Great October Socialist Revolution; the Great Fatherland War; the problem of evolution of the socioeconomic formations (feudalism-capitalism-socialism); the history of the class, revolutionary, and national liberation movements; and so on.

During the postwar period, a growing number of the institutes specializing in various fields of world history were established in the Soviet Academy. Among them were the Institute of World Economy and International Relations (1956), the Institutes of Africa (1959) and Latin America (1961), the Institute of the Economy of the World Socialist System (1960), the Institute of the International Workers' Movement (1966), and the Institute for US and Canadian Studies (1967). In addition, several bilateral permanent commissions and colloquia for cooperation between Soviet scholars and their colleagues from the socialist and

Western countries were organized within the Soviet Academy. All of these institutions contributed to expanding contacts of Soviet academia with the broader community in the internationally growing field of economic history.

All of this, in turn, led to an increase in scholarly publishing. Before Stalin's death, there were only a few historical scholarly journals including the most important one *Voprosy Istorii* [*Problems of History*]. By 1962 there were 81 Soviet periodicals devoted to history (Markwick, 2001, 67). Among the new ones, *Voprosy istorii KPSS* [*Problems of Communist Party History*], *Istoriya SSSR* [*History of USSR*], *Novaya i noveishaya istoriya* [*Modern and Contemporary History*], and *Mirovaia ekonomika i mezhdunarodnye otnoshenia* [*World Economy and International Relations*] were the most important. Some of them opened their pages to public discussions over certain methodological aspects of national, Soviet, and world history.

Soviet academic nomenklatura considered it necessary to respond to the challenges and new trends in global historiography. Soviet scholars were represented at all the international congresses of historical studies, beginning with the 10th Congress of Historical Sciences in Rome in 1955 (Filatov, 2000; Mokhnacheva, 1996). They also participated in the international forums of economic historians from 1960s. During the third such forum, which was held in Munich in 1965, two representatives of Soviet academia were included in the International Association of Economic History. The association organized its fifth conference in Leningrad in 1970; thenceforth, it began to be referred to as a congress, which assembled every four years in various countries. In 1964 the section of economic history was established within the framework of the National Committee of Soviet historians.

For Soviet historians, travel outside of the Soviet borders, in particular to Western countries, was a privilege, accessible to the rare few. It was usually utilized by those scholars who were allowed to enter the half-closed world of Soviet nomenklatura. The price for privileged status was not so much a vocal gratitude and silent humility, but an aggressive stance with regard to real or imaginary enemies of the Party. For this reason, scholars who went abroad were obliged to be not simply professionals, but also fighters on the ideological front. As a rule, the texts of their lectures were subject to rigorous inspection long before the trip and were repeatedly checked and re-adjusted.

The institutional network of historical science in the national republics was based, as a rule, on the all-Soviet pattern. The republican academies of sciences became truncated copies of the Soviet Academy. According to the division of labor between the republican and all-Union academic centers, the economic histories of the national republics were studied mainly by local researchers, both historians and economists. However, little is known about how the mechanism of their cooperation with the metropolitan scholars worked in practice. One can only note that the strategic problems that belonged to the field of methodology, interpretation of world history, and the history of the USSR as a whole remained within the competence of the central historical institutions. Republican scholars usually contacted the outside world through the mediation of the center. It is no wonder that historical writing in the national republics bore the distinct stamp of provincialism.

Ukrainian Soviet historiography followed the development of the all-Union/Russian historical sciences, although with substantial differences. Contrary to Russian historiography, it experienced several ruptures on the institutional and generational levels after 1917. Periodic campaigns against "Ukrainian bourgeois nationalism" accompanied by repression had bled Ukrainian academia. For example, of the 44 Ukrainian Soviet scholars who studied economic history in the late 1920s, only twelve survived Stalinist repressions (Danko, 1962, 62). The Ukrainian historian usually was under more tough supervision and had fewer resources and opportunities to communicate with the outer world compared to his or her colleagues in Moscow or Leningrad. Paradoxically, a reader in the Lenin scientific library in Moscow had no restrictions on access to literature on Ukrainian topics, which was forbidden to the same reader in Ukrainian-based libraries.

The process of restoration of the tradition of academic scholarship in Ukraine after Stalin's death was much slower than in Russia. The era of political thaw in Kyiv started later and ended earlier than in Moscow. Only in 1957, after a 25-year break, were Ukrainian historians allowed to establish the republican *Ukraïns'kyi istorychnyi zhurnal* [*Ukrainian Historical Journal*] as well as several university departments of Ukrainian history. The journal, however, remained under the dual jurisdiction of the Institute of History of the Ukrainian Academy and the republican Institute of the History of the Party. The latter provided the lion's share of the articles published in the journal.

The highly selective rehabilitation of pre-Soviet intellectual heritage in Ukraine did not affect either the greatest Ukrainian historian, Mykhailo Hrushevsky (1866–1934) or the leader of the Ukrainian Marxists of the 1920s, Matviy Yavorsky (1885–1937), called "Ukrainian Pokrovsky." Mykhailo/Mikhail Volobuyev (1903–1972), a leading Ukrainian Soviet economist of the 1920s who stressed that Ukraine was an integral economic complex within the Russian Empire and, as such, Ukraine remained a colony of the imperial center, was also not rehabilitated, even though he survived the purges and continued to work in the USSR after World War II.

In unison with metropolitan scholars, Ukrainian scholars were to follow the Marxist-Leninist vision of history based on the idea of progressive development of the socioeconomic formations designed to demonstrate the inevitability of the Great October socialist revolution of 1917. In parallel with this, all Ukrainian scholars were obliged to confirm the historical teleology of Ukrainian integration into the Russian/Soviet common space. The pompous celebration in 1954 of the 300th anniversary of the Pereyaslav Treaty between Muscovy Tsardom and the Ukrainian Cossack state became the most vivid manifestation of the imperial-like policy aimed to replace the national "Ukrainian" with the Soviet version of imperial Russian discourse of identity. As a result, Ukrainian historical discourse turned into an offshoot of the "all-Russian" one.

The Ukrainian SSR was not isolated from the outside world in the same way as, let us say, Albania. The republic was formally considered a member of many international institutions including the UN. But nowhere has the Soviet formula "national in form, Soviet in content" manifested itself as clearly as in the relations of Ukrainian SSR with the outer world. Despite its legal status, Ukraine was not in fact a subject of international relations. It remained an instrument of Soviet foreign policy. At the same time, this instrument retained its specific profile. Kiev accommodated consulates general of the eight "brother" countries: Poland, Czechoslovakia, Hungary, Romania, Bulgaria, Yugoslavia, the German Democratic Republic, and Mongolia. The first half of them had common borders with Soviet Ukraine.

Institutionally, Ukrainian historical and social sciences were subordinated to the respective institutions of the Soviet Academy based in Moscow. Ideologically, it was supervised by the republican branches of the Party and the KGB. The Institute of the History of Ukraine of the

Ukrainian Academy was supposed to be the main headquarter of the republican historians. In 1953, the institute changed its official name to the "Institute of History." Formally, the decision was motivated by a structural reorganization of the institute, which covered now not only Ukrainian history but also the history of the socialist countries of Eastern Europe. However, history of Ukraine remained the priority. During 1956–90, the number of specialists employed by the republican Institute of History grew from 61 to 165 scholars (*Instytut istoriï Ukraïny*, 2006). Among those who distinguished themselves in different fields of economic history, one can name, for example, Volodymyr Holobutsky (1904–1993), Ivan Hurzhiy (1915–1971), Olena Kompan (1916–1986), Mykola Leshchenko (1911–1991), and some other scholars. Most of them specialized in the socioeconomic history of Ukrainian lands in the Russian Empire.

In addition, the propaganda war with the collective West and especially with Ukrainian "bourgeois nationalism" remained one of the most important directions of Ukrainian Soviet historians' activities. To this end, a special department of foreign historiography studies was established within the Institute of History of the Ukrainian Academy. The Institute published a quarterly bulletin entitled *Zarubizhni vydannia pro Ukraïnu: naukovo-referatyvnyi biuleten'* [*Foreign Publications on Ukraine: A Reference Bulletin*] (1968–1975). But access to it, as to Western publications in general, was limited to individual researchers vetted by the KGB.

The Institute of Economics of the Ukrainian Academy also had among its leading scholars distinguished specialists in economic history of Ukraine, Russia, and Soviet Union, such as Pavlo Pershyn (1891–1970), Leonid Yasnopolsky (1873–1957), Oleksii Nesterenko (1904–1997), and some others. In 1947 a separate Department of the History of the People's Economy was established at the institute (Nebrat, 2016, 11). Its scholars produced series of monographs on a related topic and the three-volume *History of the People's Economy of Ukrainian SSR* (1983–1987). The history of Ukrainian economic thought was the second important direction of the research activities of the institute. From 1965 a collection of papers entitled *Istoria narodnoho hospodarstva ta ekonomichnoï dumky Ukraïns'koï SSR* [*History of the National Economy and Economic Thought of the Ukrainian SSR*] was published annually.

In 1978, the Institute of History and the Institute of Economics contributed to the newly established Institute of Social and Economic Problems of Foreign Countries of the Ukrainian Academy. The institute was part of the all-Soviet institutional network of foreign studies (Zhuk, 2017). The founding Director of the Institute, Arnold Shlepakov (1930–1996), combined his directorship with the duties of academician-secretary of the Department of Economics of the Ukrainian Academy and vice president of the Ukrainian branch of the "USSR-USA" Association. He specialized in the history of Ukrainian pre-Soviet labor immigration to North America.

Very soon, the institute acquired its distinct scholarly profile. It was informed by the specific geopolitical place of Ukraine as having common border with Poland, Czechoslovakia, Romania, and Hungary. The institute took on the role of coordinator of the research area devoted to cross-border economic cooperation with neighboring socialist countries. In the 1980s it organized a series of international conferences in Uzhgorod and Lviv, in which scholars from East European countries participated. The studies on economic activities of the Soviet Union in the Black Sea and the Near East regions as well as the history of Ukrainian diaspora in United States and Canada also contributed to the institute's republican identity. All of these topics were reflected in the institute's Russian-language quarterly, *Zarubezhnyi mir: sotsial'no-politicheskie i ekonomicheskie problemy* [*Foreign world: social-political and economic problems*].

Beyond the Ukrainian Academy, Ukrainian economic history was a subject of research in departments of history and economics that belonged to the system of higher education. In general, economic courses and research programs in this system were integrated more closely than historical ones into Soviet-Russian official discourse. Students of relevant specialties studied textbooks on the history of Russian and Soviet economics written in Russian. The only exception was the college textbook on the history of Ukrainian pre-Soviet economic development issued by Volodymyr Holobutsky in the Ukrainian language. However, it remained a limited edition.

Among the republican-based university and college professors, there were well-known specialists in the field, like Saul Borovoy (1903–1989) from Odesa, Dmytro Pokhilevych (1897–1974) from Lviv, or Serhii Krandievskii from Kharkiv, who published in 1964 the first and the only Soviet monograph on the historiography of economic history of the seventeenth–nineteenth centuries (Maidachevskii, 2009). However, many specialists in world history and economics who worked in Ukrainian-based institutions had no republican identity and considered themselves to be part of the all-Union/Russian scholarly community. Some of them would have been surprised if anybody had addressed them as "Ukrainian": common wisdom dictated that the "Ukrainian historian" definition applied only to those individuals who dealt solely with Ukrainian themes.

Hence, the problem of representation: Ukrainian-based scholars and Ukrainian-related topics were often disguised as "Soviet." There were hundreds of Soviet Ukrainian specialists, mainly in the fields of technical and natural sciences, who visited other countries annually as representatives of the Soviet Union. Most of them belonged to the applied and specialized research units which traditionally were supervised by the Soviet republican government rather than ideological structures. Compared to them, the number of historians and social scientists allowed to travel abroad remained limited.

The main task assigned to Ukrainian specialists was to supply the Soviet historical grand narrative with locally selected materials. Accordingly, Ukrainian topics on economics were treated as part of Russian imperial and Soviet, not national, history. Few Ukrainian historians were granted the privilege to represent the Soviet Union at international congresses of historical sciences or forums of economic historians. Even fewer of them were involved in the all-Union theoretical discussions on such topics as the transition from one socioeconomic formation to another or the development of capitalism in the Russian Empire.

At the all-Union level, Ukrainian specialists were better represented in the field of agrarian history. Studies in agrarian history were coordinated by the Research Council on agrarian history at the Soviet Academy since the end of the 1950s. Among the leaders in this field, the following can be named: Victor Yatsunsky (1893–1966), Ivan Kovalchenko (1923–1995), Andrei Anfimov (1916–1995), and Nikolai Druzhynin (1886–1986). From 1958, annual symposiums on the problems of the agrarian history of Eastern Europe were held in the USSR. The participants included scholars from most of the Soviet republics and other socialist countries. The materials from these conferences were regularly published in a special collection, *Ezhegodnik po agrarnoi istorii Vostochnoi Evropy* [*Yearbook of the Agrarian History of Eastern Europe*]. One of those collections was prepared by the Ukrainian Academy in 1962.

However, the Soviet Union never had a monopoly on Ukrainian studies. Numerous Ukrainian refugees and members of the Ukrainian diaspora in Europe and America worked in the West on Ukrainian topics. After World War II, in the course of 1945–46, Ukrainian refugees managed to re-establish a global network of Ukrainian scholarly and educational institutions. This network included the Ukrainian Free University in Munich (established in 1921 in Prague and transferred in 1945) as well as the Ukrainian Technical and Husbandry Institute (established in 1932 in Poděbrady and moved to New York in 1954), the Ukrainian Higher School of Economics (established in 1945), the Shevchenko Scientific Society (revived in 1947), and the Ukrainian Free Academy of Sciences (established in 1945,with an economic section from 1947); the last two institutions established their respective branches and offices in the United States, Canada, France, and West Germany. Scattered throughout the world, Ukrainian intellectual émigrés maintained an important degree of unity regardless of their country of residence. This transnational network informed the world about Ukrainian and Soviet-related issues from the national perspective, while also promoting Ukrainian studies as a respectable and viable field of research.

With the establishment of the Harvard Ukrainian Research Institute (1973) and the Canadian Institute of Ukrainian Studies at the University of Alberta (1976), new stimulus arose for catching up with main trends in Western humanities and social sciences and updating the Ukrainian national grand narrative. Both émigrés and those who grew in the host countries were among the specialists whose scholarly activity was connected with Ukrainian economic history. Among them, the economists Peter Woroby (1920–2002), Iwan Koropeckyj (1921–2012), Vsevolod Holubnychy (1928–1977), Zinowij Melnyk (1928–2021), and Volodymyr Bandera (1932) held professorial positions at various American and Canadian universities. Among the specialists in other disciplines, historians and social scientists Roman Szporluk (1933), Bohdan Krawchenko (1946), John-Paul Himka (1949), Wsevolod Isajiw (1933), and Peter Potichnyj (1930) addressed the economic history of Ukraine from their respective disciplinary perspectives.

The history of Ukrainian economy and economic thought was initiated and conducted by professional economists rather than by historians per se. Iwan Koropeckyj from Temple University started systematic research on the history of the Ukrainian Soviet economy. He organized a series of international scholarly conferences, one of which (in 1985) dealt with the history of the Ukrainian economy from the time of Kyiv Rus' to the Great War (Koropeckyj, 1991). Koropeckyj's exploration of Ukrainian economic thought was an important addendum to the overview of Ukrainian economic studies published earlier by his colleague, Bohdan

Wynar (1926–2013), in Ukrainian (Wynar, 1967). Each of the "two Ukraines" on the different sides of the Iron Curtain claimed to be the only true representative of Ukraine. They were keenly aware of each other and even mirrored each other to some degree. Each of them was involved in the political, ideological, and informational confrontation between the two superpowers—the United States and the USSR. From the Western side, some Ukrainian intellectuals, scholars, and social activists cooperated closely with organizations financed by CIA. From the Eastern side, historians and social scientists were obliged to be active fighters on the ideological front under KGB supervision. Contacts with the West were useful in organizing subversive information campaigns in Western countries, infiltrating emigrant institutions, and supporting communist organizations and leftist movements.

Soviet Ukrainian authorities maintained contacts with the Ukrainian diaspora through the Ukrainian Society for Friendship and Cultural Relations with Foreign Countries and the Association for Cultural Relations with Ukrainians Living Abroad (the Society Ukraina). There were also other established channels of communication with the outer world through the Academy, as well as the largest Ukrainian universities, located in Kyiv, Lviv, Uzhhorod, Chernivtsi, Odesa, and Kharkiv. Needless to say, functionaries of the Party and the KGB controlled and coordinated all of these as well as any other direct contacts of Soviet institutions and scholars with the outer world. Ukrainian left-oriented communities in Canada attracted close attention of the Soviet Ukrainian authorities. The Association of United Ukrainian Canadians (known by the Ukrainian acronym TOUK) under the leadership of Ukrainian national-communist Peter/Petro Krawchuk (1911–1997) was one of the most active participants of various exchange programs with Soviet Ukraine. Many representatives of the Soviet Ukrainian nomenklatura visited Canada to promote the official Soviet narrative. Soviet Ukraine's contacts with the Ukrainian diaspora in practice had no academic component, especially not within the politically sensitive fields of the humanities and social sciences.

Ukrainian historians on both sides of the Iron Curtain were compelled to maintain contacts by using the framework of academic exchanges established between the Soviet Union and Western countries. The framework included participation in international forums as well as various forms of scholarly exchange at the institutional and personal levels. The U.S.-Soviet General Exchanges Agreement was signed in 1958. Initially, it was coordinated by the Inter-University Committee on Travel Grants (IUCTG). In

1968, the committee was absorbed into the International Research and Exchanges Board (IREX) headquartered in New York. Western specialists were given the opportunity to travel to the USSR for work in the archives and libraries (Engerman, 2009; Richmond, 2003). Among American graduate students and junior faculty, historians were well represented, at a time when Soviet functionaries displayed an attitude of suspicion toward economists, sometimes rejecting almost half of their applications (Engerman, 2009, 170, 247). The movement in the opposite direction, from the Soviet Union to the West, was far less intensive. Within this movement, the Ukrainian presence was hardly visible at all.

Close Encounters of the Third Kind

In what follows, I will, first, describe the participation of Ukrainian scholars at the world conferences of historians; second, focus upon some aspects of academic exchange between Soviet and non-Soviet Ukrainian scholars; and third, touch on the exchange of ideas.

Diaspora Meets Home Historiography

Ukrainian historians were represented at most of the world congresses of historical sciences as members of various national delegations other than Ukrainian: Soviet, Polish, American, Canadian, and so on. The system of selection of Ukrainian participants from the Soviet side remains underexplored. There is no doubt, however, that it was guided by ideological and political rather than academic considerations. The level of representation of Ukrainian Soviet scholars was kept to a minimum and was decorative rather than substantive. The only Ukrainian Soviet scholar who participated in the 10th International Congress of Historical Sciences in Rome in 1955 as part of the Soviet delegation was Ivan Boyko (1899–1971), former chair of the Department of History of the Soviet Union at the Party's Higher School in Kyiv and chair of the Department of the History of Feudalism of the Academy's Institute of History. Volodymyr Holobutsky was the only representative of Soviet Ukraine to the 11th International Congress of Historical Sciences in Stockholm in 1960. Only the director of the Institute of History of the Ukrainian Academy, Kuz'ma Dubyna (1894–1960), was included in the Soviet delegation to the 12th International Congress of Historical Sciences in Vienna in 1965.

Soviet Ukraine was represented by as many as 56 scholars at the next International Congress of Historical Sciences in Moscow in 1970. But their presentations were not even included in the official program. At all other international congresses of a similar caliber, Ukraine was usually represented by a single scholar, most often from the ranks of republican nomenklatura. For example, Arnold Shlepakov (1930), director of the Institute of Social and Economic Problems of Foreign Countries, participated in the 1970 International Congress on Economic History in Leningrad as a "Soviet" not a "Ukrainian" scholar. Ihor Kamenetsky (1927–2008), a Ukrainian diaspora historian and participant in the Vienna International Congress of Historical Sciences, asked a rhetorical question: "Why does the Kremlin, which does not hesitate to send Ukrainian Soviet dancers and athletes abroad to various international performances and Olympiads, keeps Ukrainian Soviet historians behind the Iron Curtain? Is the main reason the underdevelopment of the historical sciences in Soviet Ukraine or the Party's fear that Ukrainian Soviet historians might publicly take the same independent positions on certain historical facts as some of their colleagues from the satellite states do?" (quoted in: Atamanenko, 2009, 265).

Ukrainian diaspora scholars were much more active in attending international forums of historians even though they officially represented their respective countries of residence. As a rule, they used every opportunity to inform the international communities of scholars on Ukrainian non-Soviet academia and its scholarly activities. For example, Mykola/Nicholas Chubaty (1889–1975), a specialist in early Ukrainian history and president of the Shevchenko Scientific Society in the United States, attended the tenth congress in Rome (1955), where he argued with Soviet historians about the national dimension of the Soviet historical writing and the history of Kyiv Rus.

Ivan Lysiak-Rudnytsky (1919–1984), one of the leading Ukrainian historians in the West, took part in several international congresses of historical sciences in 1960–70. He vividly described his impressions of meeting with Kuz'ma Dubyna, the only official representative of Soviet Ukraine at the congress in Vienna (1965) (Lysiak-Rudnytsky, 1994, 428–429). Lysiak-Rudnytsky observed this representative of Soviet Ukrainian nomenklatura with great curiosity. He understood Dubyna's privilege status as a high-ranking administrator rather than as a theoretical scholar, so he did not expect to hear from Dubyna any interpretations of the past other than official. What impressed Lysiak-Rudnytsky was Dubyna's confident and relaxed manners as well as the fact that the Ukrainian Soviet "red noble" appeared to be well informed on Ukrainian studies abroad. No wonder Lysiak-Rudnytsky became optimistic about the possibility of establishing a dialogue with the Soviet Ukrainian grandees on the common-sense ground. What he did not take into account was that Soviet citizens of that kind were used to behaving differently at home and abroad.

When Lysiak-Rudnytsky visited Moscow in 1970 to take part in the 13th International Congress of Historical Sciences, he ran into Ukrainian Soviet scholars from the Ukrainian Institute of History who discussed a topic in Ukrainian on the sidelines of the congress. One of them was Ivan Hurzhiy, a high-ranking specialist in Ukrainian economic history and representative of Ukraine on the National Committee of Soviet Historians. His companion was another well-known specialist in Ukrainian modern history, Vitalii Sarbey (1928–1999). When Lysiak-Rudnytsky joined the conversation and invited Hurzhiy for a separate meeting, the latter politely refused. Sarbey would certainly remember this episode: two decades later, he made Lysiak-Rudnytsky's ideas on Ukrainian modern history the main point of departure in his own transition from the Soviet to the national paradigm of historical writing.

All leading Ukrainian historians working for the Ukrainian Academy had access to the forbidden fruit of the émigrés' intellectual products. They were supposed to react to them publicly. During the Soviet political thaw, a sort of a dialogue between the "two Ukraines" appeared. Each side not only closely monitored the other side's scholarship but also tried to react to it. For example, the journal *Ukraïns'kyi istoryk* [*The Ukrainian Historian*], published by the Ukrainian Historical Association in the United States since 1963, was initiated in response to the appearance of the *Ukraïns'kyi istorychnyi zhurnal* [*Ukrainian Historical Journal*] established in 1957. And to some extent the *Ukraïns'ka Radians'ka Entsyklopediia* [*Ukrainian Soviet Encyclopedia*] was a response to Volodymyr Kubijovyč's multivolume *Entsyklopediia ukraïnoznavstva* [*Encyclopedia of Ukrainian Studies*], which was published in exile in various versions since 1949. After publication of the English-language edition of Kubijovyč's *Encyclopedia* in 1963, the Ukrainian Academy had the seventeenth volume of its *Ukrainian Soviet Encyclopedia* promptly updated and translated into English in 1969. Both of them contained a vast amount of information on the history of Ukrainian economics.

Ukrainian scholars in the West welcomed the political thaw in the USSR, the revival of academic historiography, and the expansion of contacts with Soviet Ukraine with optimism and hope. The establishing of new contacts between the Soviet Union and the outside world gave hope for a direct dialogue between the "two Ukraines." This concerned not only those who had relatives in the Soviet Union or agreed to participate in a program of rehabilitation. The Ukrainian diaspora faced the question of whether to establish contacts with the Soviet side, knowing in advance that they would be closely watched by the KGB, or to avoid them completely and remain in the struggle against the Soviet regime.

The most prominent Ukrainian émigré scholars, the so-called *realitetnyky* [realists], were in favor of expanding contacts with the Soviet side, confident that the Soviet regime would gradually evolve toward democracy and national freedom (Hrytsak, 2004, 114–125). Some of them, including the historian, Ivan Lysiak-Rudnytsky and the economists Iwan Koropeckyj and Vsevolod Holubnychy, even started working with Soviet Ukrainian officials on a project of a lecture tour for a group of diaspora scholars in Ukraine in 1971. However, the project was buried with the end of the political thaw in Soviet Ukraine in the next year.

Some Ukrainian historians from the United States and Canada took the opportunity to visit Ukraine. However, not everyone was allowed to do so. The Soviet side carefully filtered candidates for travel. According to historian Zenon Kohut, his personal experience with the IREX program revealed "an important point: the ability of the Soviets to shape the field of Russian and Soviet studies in the United States by determining who went and what topics could be studied" (Kravchenko, 2017, 32). From the Soviet perspective, all of those involved in Ukrainian studies in the West were considered "nationalists," controlled by or connected with American secret services, specifically the CIA. Those who managed to get to the USSR remained under the close supervision of the KGB. It was not uncommon for American students to be accused of espionage and exiled beyond the borders of the USSR.

The Soviet Ukrainian KGB archive in Kyiv contains alarming reports about Ukrainians coming from abroad. One report mentions "an American of Ukrainian descent Himka, suspected of involvement in US intelligence," who was "trying to gain access to the archives of the library" of the Academy, the Institute of Language and Literature, the Institute of History, and Kiev State University. Economist Iwan Koropeckyj, who visited the Soviet Union, also appeared in KGB memos as a secret agent.

Manoly Lupul (1927–2019), founding director of the Canadian Institute of Ukrainian Studies, was detained by the KGB near the village in the Bukovyna region where his ancestors came from, just for photographing a bridge over a local river. Andrew Gregorovich (1935–2022), the editor of the Toronto-based non-political cultural magazine *Forum: A Ukrainian Review*, described his visit to the Institute of History of the Ukrainian Academy during the Soviet political thaw. He met with Volodymyr Holobutsky and some other scholars who greeted him warmly but were not able to speak freely for fear of microphones or being observed. In private conversations, Ukrainian intellectuals asked their Western visitors not to "decipher" their work and "not to point a finger at people" (Lysiak-Rudnytsky, 2019, 641–642).

The expansion of personal and scholarly contacts between the Ukrainian diaspora and Soviet Ukrainians disillusioned those left-wing intellectuals who believed in the future of Ukraine within the Soviet Union. They were disappointed by the reality that they encountered. A vivid demonstration of this is the revelations shared with readers by the former activist of the Canadian Communist Party, John Kolasky (1915–1997) (Kolasky, 1970). In his opinion, the Soviet Union was carrying out a policy of systematic Russification, which was ruinous for Ukrainian culture. Other compatriots of his garnered similar impressions of Soviet Ukraine. As a result, the pro-Soviet segment in the Ukrainian diaspora rapidly lost its influence. Not only youth but also older activists grew disillusioned with Soviet nationality policy.

Exchange programs between East and West had different outcomes in Russian and Ukrainian contexts. Those involved in American studies of the Russian past greatly benefited from the contacts with their Soviet-Russian colleagues (Engerman, 2009, 153). Many American historians gratefully mentioned the assistance provided to them in Moscow by Petr Zayonchkovsky (1904–1983), Solomon Lurie (1891–1964), Natalia Pirumova (1923–1997), and Boris Itenberg (1921–2016). In the Ukrainian case, Soviet Ukrainian scholars tried to avoid all personal contacts with their colleagues from abroad, which could only get both of them into trouble. What was allowed in the capital turned out to be suspicious in the province.

Marxism Meets Revisionism

Two parallel processes—de-Stalinization in the Soviet bloc and decolonization in the Third World—created a situation conducive to the process of revision of the dominant historical narratives on the both sides of the Iron Curtain. They brought to the academic mainstream three interconnected themes: Marxism, nationalism, and modernization. Each of them, in one way or another, became a subject of public and scholarly discussions fueled by the iconoclastic spirit of the new generation of the 1960s, which found its way to leftist ideologies and movements. Potentially, each of the themes mentioned above could have established a meeting place for professional debates. However, it never came to that.

In the Soviet Union, the political thaw that ensued following Stalin's death was accompanied by searches for ideological, theoretical alternatives to Stalinism in the works of Marx, Lenin, and his Soviet followers of the 1920s. During the second half of the 1950s through the 1960s, a 39-volume collection of the works of Marx and Engels was published in the USSR, as were the complete collection of the works of Lenin and the works on the history of Russia by his comrade-in-arms, the first official Soviet historiographer, Mikhail Pokrovsky (1868–1932), rehabilitated posthumously. Pokrovsky's works also inspired great interest in the West when they were published in English translation. Thousands of the other texts from which a taboo was removed were returned to scholarly usage. For the most part, their authors belonged to the epoch of the Russian Empire and the first Soviet decade. The former inspired Russian nationalists, the latter attracted communist romantics. In the 1960s the works of pre-Soviet, early Soviet, and Western historians and economists began to be published in the USSR and the history of historical and economic thought was brought to the fore. Usually, the views of the non-orthodox authors were subjects of criticism. However, in criticizing their concepts and views, Soviet authors substantially broadened their mental and professional horizons.

Such exercises created fertile ground for the appearance of intellectual "heretics," who initiated independent searches in the field of Marxist philosophy and methodology, thus encroaching on the ideological monopoly of the Party structures (Krom, 2015, 60). Mikhail Ya. Gefter's (1918–1995) seminars held by the section of historical methodology of the Institute of History of the Soviet Academy in the 1960s were one example of such development. Aaron Ya. Gurevich (1924–2006), who popularized the Annales school of historical writing and came to question some Marxist theses about the origin of feudalism, was another. Both of these historians were expelled from the Soviet Academy in the 1970s. But they were not alone—intellectual challengers manifested themselves in other disciplines and centers of the Soviet Union if only in a moderate way.

In the field of socioeconomic history of the Russian Empire the new, revisionist-minded historians started discussions on the multistructured character of Russian imperial economics and the role of the state in its development. Among them were Arkadii Sidorov (1900–1966), Pavel Volobuev (1923–1997), Konstantin Tarnovskii (1921–1987), Andrei Anfimov (1916–1995), Viktor Danilov (1925–2004), and Aron Avrekh (1915–1988)—"historians who would form the core of the so-called New Direction in the late 1950s–60s, challenging the Short Course interpretation of Russian capitalism and the origins of the revolution" (Kozlov, 387–388). Discussions on the topics of Russian capitalist development proposed different answers to the question of the level of economic development of the Russian Empire prior to 1917 and therefore the level of its preparedness for the socialist revolution. New trends in Soviet historiography attracted the attention of Western researchers. Changes in Soviet policies and reforms, whose goal was to breathe new life into the Soviet project, stimulated the emergence of a left-oriented "revisionist" trend in American historical writing under the leadership of Stephen Cohen, Moshe Lewin, Sheila Fitzpatrick, and others.

The revisionists set a goal of proving, first of all, the organic character of the October revolution of 1917 and its support from below. Secondly, the revisionists were convinced that between Lenin's policies in the 1920s and the Stalinism of the succeeding decade there existed a fundamental difference, which allowed continuing discussions over possible alternatives to a Western model of capitalism. All these made American "revisionists" natural partners of their Soviet counterparts, or so it would seem. However, a partnership between the two became impossible. Even leading Soviet historians like Alexander Shapiro (1908–1994) were more interested in facts and historical sources rather than in new interpretations (Kovpak, 2014). The only opportunity for collaboration between Soviet and Western specialists was in the application of quantitative methods in historical research. The leader of Soviet cliometrics, Ivan Kovalchenko, established a laboratory for the application of computer programs in agrarian history at Moscow State University (Borodkin, 2016, 200). Quantitative economic history became a field of cooperation between the Soviet Academy and some American cliometricians. Their joint projects were realized in the final years of the USSR's existence.

As for Ukrainian economic history, it included a number of topics that had the potential for international cooperation of scholars in a broad chronological range from the earliest to recent history. Ukrainian historical-economic studies contained abundant material about Ukraine's connections with other peoples and countries. They also were important for exploring mutual relations between imperial centers and peripheries. They could be useful in comparing national and regional aspects of modernization. However, all these prospective directions of historical research were only partially claimed by the Soviet and Western scholars. Among Ukrainian Soviet historians, the revisionist trend manifested itself only sporadically, on a personal level. The unorthodox Ukrainian historians focused predominantly on issues related to the national narrative, which had nothing to do with Russian Soviet "revisionists." Volodymyr Holobutsky was the only Ukrainian historian who was invited to the all-Union debates on the transition from feudalism to capitalism. But his opinion on the early emergence of capitalism in Ukrainian economic development had no visible effect on the outcome of the debates. The majority of those who tested the boundaries of the official canon of Ukrainian historical writing turned to the history of Kyiv Rus' and the Pereyaslav agreement of 1654 (Mykhailo Braichevsky) or Ukrainian historiographical legacy (Fedir Shevchenko). The issue of Ukraine's colonial status in the Russian Empire was raised by Olena Luhova, but the discussion on this topic was stopped at the beginning.

Nationalism Meets Marxism

The issue of economic backwardness, which occupied the minds of many historians from East European and Western countries (Alexander Gerschenkron among many others), was interpreted by their Ukrainian Soviet counterparts almost exclusively through the lenses of the imperial "Ukrainian-Russian unity" paradigm rather than in terms of Marxist doctrine. The issue of modernization was explored predominantly in terms of the necessity and "inevitability" of the integration of Ukrainian economics into the all-Russian "common market."

The possibilities for Ukrainian scholars to cooperate with their counterparts beyond the Soviet borders were limited. Ukraine had shared history with the neighboring states and peoples: Polish, Hungarian, Romanian, and Czechoslovak. However, Ukrainian Soviet historians' research agenda was subordinated to the needs of ideology and current politics. It was limited to highlighting bilateral historical connections between the "brother peoples" within the discourse of their "common struggle" against capitalist "exploiters" and Nazis. Ukrainian economists, in turn, focused exclusively on the practical tasks of strengthening their socialist neighbors' economic cooperation with the Soviet Union.

The history of Ukrainian labor migration to North America could have been another possibility for scholarly cooperation between the early generation of Soviet Ukrainian Americanists and Western scholars. Arnold Shlepakov, who became a leading Soviet specialist in this field, had the opportunity to collect materials for his research during a long-term scholarly trip to the United States. "As a result, by the end of the 1970s the special field of Ukrainian-Canadian studies had been created in Soviet Ukraine, which had no equivalent in Canadian studies in Moscow" (Zhuk, 2017, 246). However, Soviet Ukrainian scholars were allowed to cooperate within this field only with the Ukrainian-Canadian left, like the communist journalist Peter/Petro Krawchuk.

It is ironic that the search for Marxist interpretations of Ukrainian national history found fertile ground in Canada, not in the Soviet Union. Bohdan Krawchenko explored socioeconomic changes in Ukrainian modern history and their impact on the national development in the twentieth century, while John-Paul Himka's research focused on the Ukrainian peasantry of Galicia and its involvement in the nascent national movement. Krawchenko made extensive use of Soviet Ukrainian historiography, while Himka was the first historian to introduce Miroslav Hroch's analysis of national movements among the so-called small nations of Eastern Europe to Ukrainian studies.

In the United States, Roman Szporluk's exploration of the mutual relations between Marxism and nationalism may be considered one of the most remarkable contributions to the growing field of Ukrainian studies in the West (Szporluk, 1988). He researched it, turning to the intellectual legacy of Karl Marx, Friedrich List, Tomas Masaryk, and Mikhail Pokrovsky. Szporluk subscribed to Gellner's interpretation of nationalism as a strategy of modernization and the centrality of the nation-state in this process; he also advocated the "national roads to modernity" paradigm

(Szporluk, 1990). In polemics with Miroslav Hroch, Szporluk criticized "a pious myth, of which Marx was the originator, that the proletariat was in its essence a supra-national class. ... Even Marxist scholars admit these days that the proletariat was never international and that it certainly had not been so before it became 'nationalized'" (Szporluk, 1990, 150).

Szporluk demonstrated on the basis of the Ukrainian example that Soviet modernization policy, which many in the West considered successful, had acquired a distinct Russian national dimension. "This leads him to hint at another definitional point of which we will hear more: states are not always quite so unitary and homogeneous as Gellner imagined" (Hall, 1998, 5). Szporluk concluded that the Soviet/Russian policy of modernization did not erase the national traditions of the peoples of the USSR, Ukrainians first and foremost, as the communists desired. This conclusion, enunciated on the eve of the fall of the communist system and the disintegration of the USSR into nation-states, proved prophetic.

Polemics of that kind were impossible in Soviet Ukraine. Instead, unrelenting criticism of "bourgeois falsifiers" remained a mandatory component of the Soviet historiographical canon and of professional training. Authors who were involved in the "critique of bourgeois forgers" industry developed a special way of writing about the texts of Western scholars regardless of the latter's methodological background. As a rule, Soviet scholars used quotations taken out of context, which made Western research an easy target for the Soviet counter-propagandists. It was desirable and sometime even mandatory to avoid any references to the sources or even to the titles of the criticized texts. At the same time, references to the Soviet canonical texts had to be accurate. Such manipulations were reminiscent of medieval scholasticism, in which polemicists beat their opponents with quotations from the Scriptures and church authorities.

The language of such publications was also appropriate. The names of Ukrainian diaspora institutions, the titulation of Ukrainian professors, sometime even their Ukrainian identification—all of this was necessarily presented by the Soviet writers in quotation marks in order to underscore their artificial, "false" character. Such methods were supposed to create in the reader the feeling that the "Ukrainianness" of the emigrants was fake, and the real Ukraine was only in the USSR. The rhetoric that was used for the description of the targeted authors and their texts also pointed to its pre-secular patterns. The enemy was usually depicted as an evil person, his or her actions as cunning and treacherous, directed against the "sacred" beliefs and convictions of the "chosen" people. Any positive mention of Ukrainian Soviet historians in émigré publications could have grave consequences for the former in their homeland. After a short political and intellectual thaw in the 1960s, Ukrainian Soviet historiography entered a prolonged period of stagnation, which lasted until the end of the 1980s.

Conclusion

In general, the study of the economic history in the USSR and in the West unfolded as if in opposite directions. Soviet scholars were interested primarily in narrow, local problems, while Western scholars focused on long-term economic processes and worldwide historic-comparative studies. This difference can be understood if one remembers that Soviet scholarship remained bound and fettered by Marxist dogma and political isolation from the global research community. Being isolated from participating in novel research and scholarly discourse, not versed in other languages, and lacking access to contemporary literature in their fields, Soviet researchers and scholars sought professional satisfaction on the backstage of positivist-like factology. All attempts to revise the rigid official canon during the short period of political thaw of the 1950s and 1960s were aborted in the 1970s and mid-1980s. The era of stagnation produced several multivolume syntheses of the history of Ukraine and Ukrainian economy, which testified to the diligence of their authors. However, for the next generation of historians, they were as "useful" as any other gigantic Soviet project.

The dynamics and main directions of the development of Ukrainian Soviet and Western historiography also remained different. Ukrainian Soviet scholarship reflected, albeit with a certain delay and in a somewhat specific form, the same stages that Soviet postwar historiography went through, from the short period of revival to the long era of stagnation. In general, the average Ukrainian Soviet historian demonstrated stronger adherence to the official dogmatism and a narrower professional outlook compared to his or her colleagues in Moscow or Leningrad to say nothing of his or her Western counterparts. Western Ukrainian studies, by contrast, entered an era of institutional and intellectual development in the 1970s, which was enabled, on the one hand, by reconnecting with previous intellectual legacy(ies), and, on the other hand, by integration into the Western academic environment. If it is possible to call the history of contacts between Soviet and Western historians a failed dialogue, then

Ukrainian Soviet historians' dialogue both with their Russian and Western colleagues failed doubly.

References

Afanas'ev, Yu. N. (1996). *Sovetskaia istoriografiia*. RGGU.

Atamanenko, A. (2009). Ukraïns'ke istorychne tovarystvo ta mizhnarodni istorychni kongresy. *Naukovi zapysky Natsional'noho universytetu "Ostroz'ka Academia", 14*, 260–272.

Borodkin, L. (2016). Economic History from the Russian Empire to the Russian Federation. In F. Boldizzoni & P. Hudson (Eds.), *Routledge Handbook of Global Economic History* (pp. 195–213). Routledge, Taylor & Francis Group.

Bovykin, V. I. (1996). Milanskii congress ekonomicheskoi istorii i nasha istoriko-ekonomicheskaia nauka. In V. I. Bovykin & L. I. Borodkin (Pod red.), *Ekonomicheskaia istoriia. Obozrenie* (Vyp. I, pp. 7–27). Moskva.

Danko, J. (1962). Suchasnyi stan ukrains'koi istorychnoi nauky ta suspil'no-politychnykh dyscyplin (z konferentsii UVAN v New York'u, 1962). *Suchasnist, 12*, 58–68.

David-Fox, M., & Peteri, G. (2000). *Academia in Upheaval: Origins, Transfers, and Transformations of the Communist Academic Regime in Russia and East Central Europe*. Praeger.

Engerman, D. C. (2009). *Know Your Enemy: The Rise and Fall of America's Soviet Experts*. Oxford University Press.

Filatov, A. M. (2000). Nauchnye sviazi istorikov v gody "kholodnoi voiny": vzgliad s "drugoi storony". *Otechestvennaia istoriia, 4*, 29.

Gatrell, P., & Lewis, R. (1992). Russian and Soviet Economic History. *The Economic History Review, 45*(4), 743–754.

Hall, J. (Ed.). (1998). *The State of the Nation: Ernest Gellner and the Theory of Nationalism*. Cambridge University Press.

Hrytsak, I. (2004). *Strasti za natsionalismom: Istorychni eseï*. Krytyka.

Kolasky, J. (1970). *Two Years in Soviet Ukraine: A Canadian's Personal Account of Russian Oppression and the Growing Opposition*. Peter Martin.

Koropeckyj, I. S. (Ed.). (1991). *Ukrainian Economic History*. CIUS Press.

Kovpak, D. (2014). Sovetskii istorik i zarubezhnaia nauka: Aleksandr Lvovich Shapiro i zapadnye istoriki-rusisty. *Forum noveishei vostochnoevropeiskoi istorii i kultury, No. 1*. http://www1.ku-eichstaett.de/ZIMOS/forum/inhaltruss21.html

Kozlov, D. (2011). Athens and Apocalypse: Writing History in Soviet Russia. In A. Schneider & D. Woolf (Eds.), *The Oxford History of Historical Writing: Volume 5: Historical Writing Since 1945* (pp. 375–398). Oxford University Press.

Kravchenko, V. (Ed.). (2017). *Ukrainian Studies in Canada: Texts and Contexts. Proceedings of the CIUS Fortieth Anniversary Conference, 14–15 October 2016*. CIUS Press.

Krom, M. (2015). From the Center to the Margin: The Fate of Marxism in Contemporary Russian Historiography. In Q. Edward Wang & G. G. Iggers (Eds.), *Marxist Historiographies: A Global Perspective* (pp. 59–71). Taylor & Francis Group.

Lysiak-Rudnytsky, I. (1994). *Istorychni ese* (Vol. 2). Osnovy.

Lysiak-Rudnytsky, I. (2019). *Shchodennyky*. Dukh i Litera.

Maidachevskii, D. Y. (2009). Ekonomicheskaia istoriografia: istoria formirovania i uroki razvitia predmetnoi oblasti. *Istoriko-ekonomicheskie issledovania, 10*(3), 5–40.

Markwick, R. (2001). *Rewriting History in Soviet Russia: The Politics of Revisionist Historiography, 1956–1974*. Houndmills.

Mokhnacheva, M. P. (1996). Sovetskaia istoricheskaia nauka na mezhdunarodnykh nauchnykh forumakh: istoki nesostoiavshegosia dialoga. In *Sovetskaia istoriografiia* (pp. 78–108). RGGU Press.

Nebrat, V. V. (2016). Istoryko-ekonomichni doslidzhennia v Instytuti ekonomiky AN URSR. *Istoria narodnoho hospodarstva ta ekonomichnoi dumky Ukrainy, 49*, 9–26.

Richmond, Y. (2003). *Cultural Exchange and the Cold War: Raising the Iron Curtain*. Penn State University Press.

Smolii, V. (Ed.). (2006). *Instytut istorii Ukrainy NAN Ukrainy (1936–2006)*. Instytut Istorii Ukrainy Press.

Szporluk, R. (1988). *Communism and Nationalism: Karl Marx Versus Friedrich List*. Oxford University Press.

Szporluk, R. (1990). In Search of the Drama of History: Or, National Roads to Modernity. *East European Politics and Society, 4*(1), 134–150.

Udod, O., & Yas', O. (2020). *Kul'turnyi prostir ukraïns'koï istoriohrafiï u svitli radianizatsiï ta socio-kul'turnykh transformatsij XX-pochatku XXI st*. In-t istoriï NANU Press.

Wynar, B. (1967). *Materiialy do istoriï ekonomichnykh doslidiv na emigratsiï*. Ukrainian Historical Association.

Yaremchuk, V. (2009). *Mynule Ukraïny v istorychnii nautsi URSR pisliastalins'koï doby*. Vyd-vo "Ostroz'ka Academia".

Zhuk, S. I. (2017). *Nikolai Bolkhovitinov and American Studies in the USSR: People's Diplomacy in the Cold War*. Lexington Books.

CHAPTER 3

East Germany: Negotiating Conformity and Innovation

Axel Fair-Schulz

The German Democratic Republic (GDR) preferred to think of itself as the "better" and "more progressive" German state, heir to all that was forward-looking, emancipatory, and edifying in German history, while West Germany became equated with everything reactionary, backward, repressive, and destructive. As the division of postwar Germany into two separate states appeared less and less provisional, and any thought of reunification become increasingly remote, the GDR changed its official rhetoric from hope for a future united socialist Germany to a newly emerging East German socialist nation. The foundation for this new and distinctive East German nation was "not language or ethnicity but class and social systems," as Georg Iggers put it (Iggers, 2017, 36). The GDR's historians played a cardinal role in fleshing out the specific traditions and legacies that state and Party drew upon in forging their new East German national identity.

A. Fair-Schulz (✉)
State University of New York, Potsdam, NY, USA
e-mail: fairsca@potsdam.edu

A. Dolezalova and C. Albrecht (eds.), *Behind the Iron Curtain*, Palgrave Studies in Economic History,
https://doi.org/10.1007/978-3-031-31578-7_3

The GDR's economic historians shared in those efforts, albeit less prominently than those engaged more directly in questions of political and ideological history. Yet, all historians had to participate in one way or another, carving out the details of how East Germany could root and legitimize itself historically. As time went on, the GDR leadership charged the state's historians with enlarging their ideological gaze. They were called upon to do justice to the past and to avoid the former schematic approach that had distinguished sharply between "progressive" and "reactionary" personalities and forces. The concept of "tradition" was now understood more broadly to comprehend not only the heroic line from Thomas Müntzer through Karl Marx and August Bebel to Karl Liebknecht, Ernst Thälmann, and finally the leaders of the GDR—and perhaps as a second thought the unnamed working poor, who in Brecht's words, built Thebes—but also those who, in paving the way for German nationhood, had created the possibility of a socialist Germany (ibid., 36).

In addition to probing Germany's past for its usefulness to the East German state, the GDR's economic historians also had to come to terms with the traditions, schools of thought, and personalities in their own field as it developed since the nineteenth century. The highly influential German Historical School played a key role in this bequeathed legacy, as epitomized by iconic figures such as Max Weber, Gustav von Schmoller, and Werner Sombart. Of course, one cannot reduce this complex and multi-faceted inheritance to those scholars and their methodological orientation alone, as Jan-Otmar Hesse has pointed out (Hesse, 2016, 97–98). Overall, the economic historians who dominated the field and university positions throughout the nineteenth and the early twentieth centuries shared the defining political and cultural leanings of academic historians in Germany by favoring nationalism, conservatism, and imperialism (For a good discussion of this see: Iggers, 1983; Ringer, 1969; Weber, 1984).

The GDR's economic historians were oriented toward historical materialism, which was a core requirement for all of academia in East Germany, especially in the humanities and social sciences. Yet, the relatively strong empirical components of economic history enabled its leading East German practitioners to resist the reductionist formulas that the ideological guidelines of the Socialist Unity Party (hereafter the Party) called for. Over the decades, they produced substantive scholarly works that combined Marxist approaches with the history of everyday life, as well as microhistory and cultural history. East German economic historians favored a more comparative approach than their West German counterparts, examining economic developments in Europe as well as in Africa, Latin America, and Asia. During the 1970s and 1980s, the GDR's economic historians consciously developed a greater sense of professional independence from the Party propagandists and focused on new areas of research, such as entrepreneurs as carriers of technological progress and environmental history (Fischer & Zschaler, 1998, 386; T. Kuczynski, 2009, 4). In addition, they began to incorporate selective insights and research questions from Western schools of thought into their own Marxist outlook, finding inspiration in such diverse approaches as cliometrics/new economic history, the Annales school, and the Bielefeld school. Jürgen Kuczynski (1904–1997) also notes Marxist scholarship from Britain, the U.S., and Japan, such as the works and ideas of Maurice Dobb, Christopher Hill, Paul Sweezy, and Kohachiro Tokahashi (J. Kuczynski, 1981, 233).

This chapter probes how different generations of East German economic historians pursued scholarly cooperation and exchange with their Western colleagues during the Cold War, as well as the obstacles and support they encountered in doing so. This requires taking stock of economic history in the GDR as an academic discipline, sketching its major stages of development, and correlating these intellectual trajectories to the changing ideological, political, and geo-strategic circumstances of the Cold War, given East Germany's location on the fault line of this gargantuan conflict.

Economic History Evolution: Historical, Political, and Ideological Background

The former GDR defined itself as a real-existing socialist state by and for workers and farmers. In the decades since that state's dissolution in 1990, even some of the GDR's most partisan defenders have come to acknowledge a significant gap between the humane and emancipatory socialist ideals on the one side, and a more contradictory and often regimented as well as oppressive reality on the other. East Germany's scholars, including its economic historians, were, by choice or default, oriented toward the officially sanctioned Marxist-Leninist framework, which included some elements of Marxist analyses but transformed Marxism from a critical perspective into an instrument of hegemony and legitimacy for the ruling regime. In addition, they inherited the intellectual and cultural traditions of their non-Marxist predecessors who dominated German scholarship until the end of World War II and the creation of the GDR on the territory of the Soviet zone in October 1949, following the establishment of the Federal Republic of Germany (FRG) on the French, British, and American occupational zones on May 23, 1949. At the same time, and more importantly, they defined themselves as Marxist scholars engaged in transforming their academic discipline. Given the Cold War context of their work in the GDR, they had to orient themselves to what successive Party leaderships in East Berlin and in Moscow defined as canonical Marxist-Leninist interpretations of Marxism at different historical moments. As the GDR was a German-speaking state at the frontline of the Cold War, East Germany's economic historians found themselves confronted not only with the task of defining and defending their brand of Marxist economic history *vis-à-vis* Western scholarship in general but West German scholarship, in particular, as both German states were locked in intense ideological competition.

Embedded in the geopolitical, ideological, cultural, and socioeconomic circumstances of the Cold War, understanding the conditions under which the GDR's economic historians operated—their opportunities, challenges, and limitations—requires addressing how the GDR defined itself and its social order, as well as the GDR's relationships with the USSR, on the one hand, and with West Germany, on the other. Those relationships conditioned the GDR's overall space for maneuver and adjustment. The GDR defined itself for most of its duration as an anti-fascist German state by and for workers and farmers. It owed its existence to the Soviet Union and could not have survived without Soviet patronage. Much like the other Warsaw Pact countries in Central and Eastern Europe, the GDR oriented itself toward the Soviet model of state socialism, which was officially referred to as "real-existing socialism."

The GDR was the product of the defeat of Nazi Germany, on the part of the USSR and its Western allies, as well as the Cold War. The division of postwar Germany into four zones of occupation, and the subsequent coalescing of the American, British, and French zones into the Federal Republic of Germany in September 1949, left the Soviets with no alternative to giving the green light, in October 1949, to the creation of the East German state. The German Democratic Republic was doubtlessly Moscow's creation, albeit an "unwanted child," as characterized by the historian and political scientist Wilfried Loth (Loth, 1998).

The USSR would have preferred a united but neutral Germany, capitalist yet indebted to the Soviet state, which demanded reparations for the horrendous destruction, war crimes, and atrocities inflicted on it by the German invaders and their allies during World War II. Stalin and his inner circle did not give the green light to the East German communists around Walter Ulbricht and Wilhelm Pieck to respond to the creation of the West German Federal Republic in May 1949, until September 1949. The GDR was officially proclaimed on October 7, 1949, yet it remained the default option from the Soviet perspective. Western separatism articulated in Chancellor Konrad Adenauer's memorable phrase "lieber das halbe *Deutschland ganz als das ganze Deutschland halb*" [better all of half of Germany rather than half of all of it] was the response to the Soviet Union's renewed offer of German unification from March 10, 1952. The Soviet proposal included the withdrawal of all Allied occupational troops from German soil within a year, political independence, free elections, and enshrined guarantees of neutrality. Adenauer's government, and the U.S. as well as its British and French allies, rejected this offer. A somewhat similar deal, however, materialized in the case of Austria in 1955, when the USSR, the U.S., Britain, and France signed the Austrian State Treaty, which allowed for full Austrian independence under conditions of neutrality until the end of the Cold War.

By 1952, when it became clear that German reunification would be impossible for the foreseeable future, East Germany embarked officially on the construction of state socialism. Prior to that time, the GDR had merely committed itself to building an anti-fascist state dominated by workers and farmers. The nature and substance of the GDR's specific socioeconomic and political system has been the subject of considerable debate during that state's existence and beyond. Different explanatory models have been proposed, reflecting various levels of analytical or polemical quality.

The newly established East German state sought a sharp break with the recent Nazi past, as well as with the even more deeply entrenched traditions of nationalism, conservatism, and imperialism in academia by actively promoting a new scholarly cohort of anti-fascist and Marxist professors. Professors with ties to the former Nazi regime were dismissed from their posts. As early as June 1945, the new president of the University of Berlin, which was located in the Soviet sector, informed his colleagues that everyone had to submit a biographical sketch and explain their relationship with the Nazi party (NSDAP) as well as overall political attitudes and activities since 1933, in addition to listing their main publications, lectures, seminars, and future plans. Membership in the NSDAP resulted in immediate dismissal in almost all cases. As of March 1945, 85 percent of all professors at the University of Berlin were NSDAP members. At the University of Jena, 98 percent of all professors were removed from their teaching positions, while at the University of Greifswald only 46 percent had to resign initially (Schneider, 1977). The latter two universities were located in the Soviet zone, where de-Nazification was pursued much more consistently than in the American, French, and British zones of occupied Germany (Steiner, 2004).

The GDR, as a self-declared democratic republic, is frequently labeled as a "communist dictatorship." This term, while widely used in both academic and journalistic circles, has little explanatory power and is actually misleading. The ruling Party defined the state's system as "real-existing socialism," to distinguish it from democratic socialist as well as the Eurocommunist projects in the West, on the one hand, and the far more advanced "communist" stage that would eventually develop out of "fully developed real-existing socialism" at some future time. The GDR was not a society where, in the words of the *Communist Manifesto*, the free development of each is the condition for the free development of all. Under East Germany's authoritarian system, the free development of the individual was limited and could not serve as a meaningful foundation for the free development of all. Nor was Marx's maxim, "communism's" defining operative principle: "from each according to his abilities, to each according to his needs," characteristic of the GDR's socioeconomic structure. Peter Grieder notes in his succinct overview of the GDR that what the Party built in East Germany "was socialism without democracy, namely, distorted socialism" (Grieder, 2012, 130).

Other scholars have advanced their own categories to describe and explain the specific social order that matured in the GDR. They range from "totalitarian welfare state" (Charles S. Maier) to "post-totalitarian party police state" (Mike Dennis), "modern dictatorship" (Jürgen Kocka), "participatory dictatorship" (Mary Fulbrook), "consensus dictatorship" (Martin Sabrow), "consultative authoritarianism" (Peter Christian Ludz), "niche society" (Günter Gaus), "welfare dictatorship" (Konrad Jarausch), "*durchherrschte Gesellschaft* [a society ruled all the way through]" (Alf Lüdke), "Stalinism" (Hermann Weber), and "Russian satrapy" (Hans-Ulrich Wehler) (Grieder, 2012, 7–16). The debates about the relative merits of those analytical categories and labels are likely to continue.

There is, however, broad agreement that the East German state was defined by the gap between humane ideals and aspirations and an at least partially repressive reality.

However, it would be overly simplistic to posit a dictatorial and "totalitarian" East in contrast with a democratic and pluralistic West. As Georg Iggers observed: "[w]estern historians have contrasted the pluralism of interpretive approaches in the West with the limits that an official interpretation of history placed in the work of the historian in the socialist countries. As it has been crudely put, historians in the West derived their findings from the evidence; those in the socialist countries fitted their evidence into their preconceived conclusions" (Iggers, 1991, 31).

Reality was and is, of course, more complex. Most tenured historians in the advanced capitalist countries of the West hold views that are largely agreeable to the dominant economic and political elites. There are, of course, some dissenters who identify as Marxists or who otherwise divert from the established consensus. However, they are few and far between, encountering greater professional obstacles than their mainstream colleagues, having greater difficulties securing funding, and finding their books frequently ignored by the corporate media (Bhambra et al., 2018; Giroux, 2014; Heller, 2016; Miliband, 2009).

The GDR was involved in an intense competitive struggle with the West German state. The FRG refused to recognize East Germany as independent and sovereign and instead referred to it semi-officially and dismissively as the "Eastern zone." Between 1955 and 1969, West Germany's official approach toward East Germany was conditioned by the so-called Hallstein Doctrine, which essentially stated that West Germany represented all of Germany, including the territory and population of East Germany. Thus, any country that established diplomatic ties with the GDR, or maintained such ties, would commit an unfriendly act. West Germany punished such "unfriendly acts" with the immediate cessation of diplomatic ties with those countries. Given that the FRG, as a beneficiary of Marshall Plan money and the "Economic Miracle" boom, became an appreciably more attractive potential partner for third countries, the GDR suffered considerable harm under these policies. East Germany, of course, reacted with outrage, rejecting West Germany's claim to represent the entire German nation. Instead, the GDR emphasized its own identity as an anti-fascist state, while the FRG positioned itself as the legal successor not only to imperial and Weimar Germany, but to the Nazi state as well.

Two of the GDR's most prominent economic historians, Jürgen Kuczynski and Hans Mottek (1910–1993), had fled Nazi Germany as communists and Jews, eventually finding refuge in Great Britain. Others, like Manfred Nussbaum (1922–1981) and Gunther Kohlmey (1913–1999), escaped to China or the Soviet Union. Kohlmey, Kuczynski, and Mottek finished their university studies before the Nazis came to power; Kohlmey and Kuczynski had earned their doctorates as well. Thus, they were exposed to, and familiar with, the main intellectual and academic trends and personalities that dominated German and international academia during the first half of the twentieth century. This equipped them with the necessary intellectual and cultural dexterity to develop a more nuanced and creative Marxism than the official Marxist-Leninist version authorized by the Party. Other influential East German historians such as Ernst Engelberg (1909–2010) and Walter Markov (1909–1993) spent years in fascist prisons.

The GDR welcomed refugees who had to escape the murderous dictatorship of the Third Reich and offered them generous academic prospects, as long they remained committed to the building of state socialism under the leadership of the Party. West German universities and colleges, on the other hand, oscillated between indifference and hostility toward refugees from the Nazi regime, especially if they had a left-leaning background. At the same time, professors who were closely associated with the Nazi regime were relatively quickly integrated into the newly established West German academic system (Schulze, 1993). For example, Walter Markov, who was imprisoned by the Nazis for ten years for his communist convictions, was denied the opportunity to submit his second dissertation (*Habilitation*) at the University of Bonn in 1946. He later moved to East Germany and had a long and distinguished career as professor at the University of Leipzig, becoming one of the leading experts on the French Revolution as well as a friend and collaborator of Albert Soboul. Markov held visiting professorships in Nigeria, Sri Lanka, and Chile, in addition to becoming the founding director of the Department for African and Near Eastern Studies. Markov and his students, the most prominent among them being Manfred Kossok (1930–1993), focused from the 1970s on comparative global history. Despite his credentials as a committed communist, anti-fascist, and Marxist scholar, Markov ran afoul of the Party leadership, was disciplined, and was temporarily demoted by the Party (Heitkamp, 2003, 110–118, 148–158).

Remarkably, Markov kept his professorship despite having been expelled from the Party in 1951. He never formally applied for re-admission even after his rehabilitation. Party membership was a fundamental requirement for all aspiring scholars in the social sciences and the humanities in the GDR. Iggers observed "[a]fter polling colleagues in the GDR, I was able to find only Wolfgang Jacobeit and Bernd Töpfer, historians and full professors at the Humboldt University who were not members of the SED or one of the parties associated with it" (Iggers, 1991, 40). The reference to the "parties associated with it" points to political parties and mass organizations connected with the Party and under its leadership, such as the Christen Democratic Union, the Liberal Democratic Party, the Democratic Farmer's Party, the National Democratic Party, the Cultural League, and the Democratic Women's League, among others.

Party membership or at least active participation in the Free German Youth Organization (FDJ) was not only key to a head-start on the academic trajectory that might one day lead to a professorship but also in many ways necessary even to be admitted to university studies in history and related fields. Refusal to join the nominally voluntary FDJ (Freie Deutsche Jugend [Free German Youth]), which essentially functioned as the Party youth organization, or for young men to submit to the required minimum one-and-a-half-year military service (as opposed to volunteering for three full years), let alone active and engaged church membership, ended most careers before they began. Karlheinz Blaschke (1927–2020) was a rarity, as an openly non-Marxist social historian in the GDR who eventually found a modest position at a Protestant Church educational institution. His institution, the Leipzig Seminary, was operated by the Lutheran Church of Saxony but not officially recognized by the GDR. It offered theological degrees for ministers. Blaschke was the only instructor there who specialized in social and regional history, as opposed to a field more directly related to theology. He refused to make the necessary ideological adjustments to several manuscripts and instead withdrew them from publication (Blaschke in Pohl, 1997, 45–93). His case illustrates how even internationally respected scholars could be marginalized and excluded from GDR academia if they refused to pay at least lip service to the official Marxist-Leninist terminology.

Economic history in the GDR was at first embedded in universities, such as at the Institute for Economic History at Humboldt University in East Berlin. There were six universities in the country, augmented by several more teachers' colleges and technical colleges. Professors at those institutions were, of course, obliged to do some research, but their main responsibility was teaching. The main institutional vehicle for research was the gargantuan GDR Academy of Science (hereafter the Academy). Between 1956 and the absorption of the GDR into the enlarged FRG on October 3, 1990, the main institutional base for East German economic history was the Institute for Economic History. This institute, closely associated with the name of its founding director Jürgen Kuczynski, was internationally recognized for its innovative scholarship and uniqueness within the GDR, given its open atmosphere and critical spirit. Here, Marxism was not reduced to an ideology that legitimized the dictatorship of the Party leadership over the rest of society but instead was productively applied and developed in critical engagement with other currents, such as historical sociology, the Annales school, and cultural history.

An important vehicle for scholarly debate, not only between East German economic historians and their Soviet bloc colleagues but also East German and Western experts in general, was the famous *Jahrbuch für Wirtschaftsgeschichte* [*Yearbook for Economic History*]. The *Yearbook* was published under the umbrella of the Institute for Economic History and included international scholars on its advisory council, such as the influential British historian Eric Hobsbawm (Fair-Schulz, 2006).

The Institute of Economic History was part of the Academy, which eventually grew to over fifty institutes and employed around 24,000 people. Another important location for economic history coalesced at the University of Economics in Berlin-Karlshorst. Created in 1950, the University of Economics was established by the government explicitly to educate experts and administrators of the socialist planning system. Economic history was only one of several fields on which teaching and research was focused. Economic geography, political economy, statistics, law, accounting, history of the working-class movement, financial planning, technology, and business administration were all part of the curriculum. Economic history in Berlin-Karlshorst was centered at the Institute for Economic History, headed by Hans Mottek, who also had an appointment as full professor of economic history and was commonly regarded as the head of the so-called Mottek school. This informal "school," which since the 1970s consisted of approximately three or four junior professors and a slightly larger number of lecturers, focused mainly on comparative economic history in Eastern Europe and the problem of economic underdevelopment in Eastern Europe relative to Western

Europe (Roesler, 2017, 147–148). From the 1970s, Mottek turned toward environmental history and played a key role in its development within East German academia (Hoell, 1989).

Encapsulated and circumscribed by the ideological guidelines formulated and enforced by the Party, the GDR's economic historians nevertheless tried to develop a more creative Marxist approach that went beyond mere partisanship in the service of Party and state. As Marxists, they were critical of the triumphalist claims of supposedly superior Western rationality and market capitalism. They balanced their training in European history and economics early on, with a growing focus on the non-Western world, not only Eastern Europe but Africa, Asia, and Latin America.

Yet, as Western observers such as Georg Iggers and GDR economic historians like Jürgen Kuczynski himself pointed out, far too many GDR scholars had internalized the corrosive norms of conformity and Party discipline, even when opportunities for greater leverage existed. Iggers notes that the GDR relaxed its tight control of historical scholarship in the 1970s, partially in order to secure broader international respect and recognition. GDR historians, in Iggers' judgment, failed to take full advantage of new possibilities, and, to cite but two examples, did not engage as readily as their Polish and Hungarian colleagues with new methodological approaches, such as the influence of mentalities and everyday culture (Iggers, 1991, 13). Kuczynski lamented that "apart from not even a dozen social and cultural scientists in the GDR, who of course fell into temporary disfavor with the party leadership," the majority neglected to engage with the most creative and fertile aspects of the Marxist tradition and instead used Marxist concepts only in a schematic, formulaic, and opportunistic fashion. "The overwhelming majority, almost all the scholars in the social sciences and the humanities ... in the countries of so-called Real Socialism [the Eastern Bloc countries] turned against [genuine] Marxist methods, forgot them or did not apply them ... They were Pseudo-Marxists" (J. Kuczynski, 1994, 266).

West Germany's refusal to recognize the GDR diplomatically, combined with the Hallstein Doctrine, overall Cold War tensions, the fact that the GDR's domestic currency was not freely convertible and thus literally valueless outside the GDR and the Eastern Bloc countries, and limited travel to conferences in the West and the acquisition of Western scholarly literature, made it difficult for East German scholars to compete. The Party apparatus, which had to approve each travel request and provide the necessary travel funds in scarce Western currency, was not generous in approving applications from within the GDR.

The GDR's geopolitical isolation improved in the 1970s as détente deescalated the Cold War and led to a normalization of ties to key Western countries. Even the FRG eventually accepted the existence of East Germany as a political reality. This furthered the space for scholarly contact and exchange. During the years between the building of the Berlin Wall in 1961 and the early 1970s, East German economic historians were rarely able to visit the West. In fact, contact with Western colleagues was largely in the hands of the two elder statesmen of economic history in the GDR, Jürgen Kuczynski and Hans Mottek (Fischer & Zschaler, 1998, 386). Both the domestic and the international Cold War context in which GDR economic historians had to operate explain why that was the case. The GDR was internationally still relatively isolated outside the Soviet bloc and parts of the Third World, largely due to West German pressure on any states that considered establishing normal diplomatic relations with East Germany. The Party leadership responded to this overall situation of contested political and diplomatic legitimacy with heightened surveillance and distrust of any personal contacts between East German scholars and their Western colleagues. Only at the end of the 1960s, as the new West German government of Willy Brandt and Walter Scheel accepted the existence of the GDR as a fact of life and moved toward de facto diplomatic recognition, did the Party leadership feel secure enough to expand the number of GDR scholars who were authorized to attend conferences in the West.

Domestically, the short period of relative political and ideological liberalization after the erection of the Berlin Wall in August 1961 was followed by a period of renewed ideological rigidity. The 11th Plenum of the ruling Party, held in December 1965, set into motion a set of more restrictive policies that were justified as a necessary defensive reaction to the growing influence of "nihilism," "skepticism," and "pornography." In addition to ideologically unreliable artists and intellectuals within the GDR, Party leaders blamed the overall corrosive influence of Western ideologies. Erich Honecker was at that time the number two in the Party hierarchy after Party leader Walter Ulbricht. Lower-ranking Party administrators were blamed for having neglected their supervisory duties and for having been too indulgent of ideological heterodoxy (Agde, 1991).

Kuczynski and Mottek belonged to the generation of economic historians that was trained and socialized prior to the creation of the GDR, and both were "Westemigranten" (refugees from the Nazi regime) who escaped to Western capitalist countries, both ending up in Great Britain during World War II. While the SED regime viewed those who returned from Western exile to live and work in the GDR with considerable suspicion, it ultimately accepted that they were already exposed to and "contaminated" by Western influences. However, as veteran communists, Kuczynski and Mottek had also proven their overall loyalty to the GDR and its political-ideological project. Kuczysnki noted in his diary on February 15, 1972, that "Hager [senior Politburo member with oversight in matters of ideology] has been told several of my remarks [criticizing various policies and decisions], who, I am told, reacted this way: "'If comrade K[uczysnki] criticizes the party then it is to be judged differently. He is an old and loyal comrade'" (J. Kuczynski, 1992, 167). Hanna Wolf, the much-feared, hardline director of the Party school at the Central Committee, wrote a letter in early 1984 to Kuczynski, chiding him for diverting from the Party line: "[y]ou are wrong, you are wrong in your usual charming ways, but you are wrong nevertheless! Yet, to be wrong is human and the good thing about you is that one can always discuss things with you (because you are wrong due to your overbearing [emphasis in the original] love of our cause and trying to change your mind is a fool's errand ...." (Wolf, 1984). The implication of Kurt Hager and Hanna Wolf's statements is clear: veteran comrade scholars like Kuczysnki had special leeway regarding independent action. Their long and proven allegiance to the Party granted them certain limited privileges, such as being able to travel freely to the West and even challenge statements by Party leaders, albeit very guardedly—a privilege that other less-seasoned comrades did not have. Kuczynski remained, of course, acutely aware of how fragile and contingent such extra latitude was. The Party leadership was less confident about the next generation of scholars who came of age in the GDR. Were they properly immunized against whatever challenges and temptations that conference trips and colleagues in the West could present?

Contacts with the West were usually managed by the Directorate for International Relations. Its decision-making process was informed, on the one hand, by practical and economic criteria and, on the other, by ideological ones. For example, the officials at the Directorate for International Relations were suspicious of overly collegial contacts between GDR and Western economists, as the latter, by definition, had to be apologists for the capitalist system. Thus, too much fraternizing between Eastern and

Western scholars could potentially undermine ideological clarity and partisanship. In addition, GDR authorities feared that East German scholars traveling to conferences in the West might, inadvertently or not, discredit the GDR by speaking too frankly about things that Western colleagues could not understand. Finally, there was the ever-present concern about confidential and strategically significant information landing in the wrong hands.

Access to Western academic literature was severely limited, for political and financial reasons. GDR scholars who personally received books or academic journals from Western colleagues and institutions were supposed to donate the material to whatever East German institute or university they were affiliated with. Yet, academic library catalogues would not normally list Western holdings. Such material was accessible only in a relatively small reading room, available to *bona-fide* scholars. Students could enter only with a special permit, issued if the material was needed for a seminar or thesis paper (Strohe, 1996, 25–27).

The Peculiarities of East German Economic History: A Window Through the Iron Curtain

The Institute for Economic History, as the premier East German hub for scholarship on its subject matter, was—especially by GDR standards—well connected to other academic institutions inside the country as well as globally. Its formal ties to Western institutions included Japanese research institutions in Kyoto and Tokyo, the University of Liverpool in the UK, as well as the Max Planck Institute for History in Göttingen and the University of Mannheim in West Germany. Other European partner universities in Amsterdam, Siena, and Stockholm were also included. In the U.S., the State University of New York (hereafter SUNY) at Binghamton functioned as the main partner institution (Fair-Schulz, 2017, 158).

SUNY Binghamton was a particularly good fit, given the role and impact there of Immanuel Wallerstein. Wallerstein, one of the main representatives of World-System's theory and a leading heterodox Marxist working on the frontiers of historical sociology and economic history, directed the Fernand Braudel Center for the Study of Economies, Historical Systems, and Civilization. Wallerstein himself taught at SUNY Binghamton as Distinguished Professor of Sociology between 1976 and his retirement in 1999. The Braudel Center was established in 1976 and, like the Institute for Economic History in East Berlin, strove to bring together a well-defined theoretical framework with detailed empirical studies. The Braudel Center's original mission statement announced in 1976 that the Fernand Braudel Center for the Study of Economies, Historical Systems, and Civilizations exists to engage in the analysis of large-scale social change over long periods of historical time. "We operate on two assumptions. One is that there is no structure that is not historical. In order to understand a structure, one must not only know its genesis and its context; one must also assume that its form and its substance are constantly evolving. The second assumption is that no sequence of events in time is structureless, that is, fortuitous. Every event occurs within existing structures and is affected by its constraints. Every event creates part of the context of future events. Of course, there are ruptures in structures which represent fundamental change. But such ruptures too are explicable in terms of the state of the structures. We, therefore, do not separate the study of historical sequence and the study of structural relationships" (Binghamton University, 2022).

Given the constraints that GDR historians had to work under, leading Western colleagues recognized their not infrequent "very substantial contributions to international scholarship, especially in less sensitive areas of study such as agrarian history and the history of the Middle Ages" (Iggers, 2017, 35). Yet, East Germany economic historians made considerable contributions to more contemporary fields and topics as well. One notable example is Dietrich Eichholtz's massive three-volume study of the German war economy during World War II, which examined in detail and on a broad empirical basis the ties between German industrialists and Nazi leaders. Eichholtz's volumes originally came out in East Berlin between 1969 and 1996. A second edition, expanded to five volumes, appeared in Munich in 1999.

The Case of the Nussbaums

Also important are the works of the Nussbaums from the 1960s. Helga (1928–2015) and Manfred Nussbaum were academic colleagues as well as marriage partners. Both worked at the Institute for Economic History; Helga eventually became its director, while Manfred worked as a section leader. In addition, both were promoted to research professorships. Manfred Nussbaum specialized in the economic history of the Weimar Republic (1919–1933), while Helga explored the connections among economic, social, and political developments in Imperial Germany (1871–1918).

Manfred Nussbaum developed expertise in another area of specialization: German colonialism in the late nineteenth century. He authored two important books on the subject matter: *Togo—Eine Musterkolonie?* [*Togo—A Model Colony?*] and *Vom "Kolonialenthusiasmus" zur Kolonialpolitik der Monopole* [*From 'Colonial Enthusiasm' to the Colonial Policies of the Monopolies*]. Both volumes came out in the early 1960s and critically examined the nexus between economic and political dynamics as they influenced Germany's quest for colonies in Africa, which gained significant momentum in the early 1880s. West German historians, such as Percy Ernst Schramm, in contrast, produced works during that time period that justified and even celebrated colonial expropriation and exploitation.

Manfred Nussbaum examined the evolving dynamics of German colonialism in Africa during the administrations of chancellors Otto von Bismarck, Leo von Caprivi, and Clodwig zu Hohenlohe-Schillingsfürst, spanning the years from 1871 until 1900. His overall approach was, of course, informed by Lenin's famous work on imperialism as a manifestation of finance capitalism. Yet, Manfred Nussbaum based his study on a broad and nuanced empirical foundation, and he deliberately avoided overly schematic interpretations and over-generalizations. In fact, he described in some detail the conflicts and debates between different wings of the ruling class, including layers of capitalists still committed to free competition and those who had fully embraced the finance capitalism stage of development. While leaving no doubt about his identity as a deeply committed Marxist historian, he labored to produce empirically informed scholarship that could not be reduced to mere propaganda or "wasted paper." His colleague and partner Helga worked for similar objectives.

Helga Nussbaum, starting as early as 1966 with her solidly grounded, empirical, and seminal book *Unternehmer gegen Monopole: Über Struktur und Aktionen antimonopolistischer bürgerlicher Gruppen zu Beginn des 20. Jahrhunderts,* contended that Marxist-Leninist assertions about how capitalists are automatically monopolistic, at the beginning of the twentieth century, did not reflect a more complex and contradictory reality.

Already in the introduction Nussbaum argued against grand and tidy generalizations, such as terms like *Grandbourgeosie*, the middle-level bourgeoisie, and small-scale entrepreneurs. She pointed out that what might be true for some groups and individuals in different areas of Germany might not be valid elsewhere. In *Wirtschaft und Staat in Deutschland: Wirtschaftsgeschichte des staatsmonopolistischen Kapitalismus in Deutschland vom Ende des 19. Jahrhunderts bis 1945* (published in 1978), she argued along similar differentiated and nuanced lines.

Despite being published only in East Berlin (albeit with the prestigious *Akademie Verlag*), Helga Nussbaum's *Unternehmer gegen Monopole* was recognized in the West as well as a solid accomplishment. George W. F. Hallgarten wrote a glowing review in the *American Historical Review* in 1967, praising the book as a "solid study," that is "a new example of the increasing realism that characterizes so many contemporary works produced by scholars behind the iron curtain, especially in the satellite states." Hallgarten commended Nussbaum for "presenting a clear and diversified picture of big German monopolies in the early years if Wilhelm II" (Hallgarten, 1967, 519). In doing so, Nussbaum offered greater nuance to Lenin's (and Bukharin's) classic studies of monopoly and finance capitalism. While Bukharin was, since his murder by the Stalinist bureaucracy, *persona non grata* and thus could not be mentioned anywhere inside the Eastern bloc, Lenin was treated as sacrosanct and his writings as canonized texts—unquestionable, unapproachable, and thus lifeless. Hallgarten was aware of Nussbaum's dilemma of having to navigate not contradicting Lenin, on the one hand, while doing justice to her diverse sources on the other. He applauded her for solving this problem pragmatically, "by presenting the great variety of forms the monopolies assumed in actual life." Hallgarten continued by contrasting Nussbaum positively with a Western colleague who echoed the standard defenses of capitalism typical for that time period. "This makes them more objective than they formerly could afford to be and often renders them superior to their Western colleagues who was a rule either ignore or blindly defend the monopolies, as does the West German economic historian Wilhelm Treue whom Frau Nussbaum justifiable criticizes" (ibid., 519).

Helga Nussbaum and Lotte Zumpe (1922–1991) were the editors of this three-volume project, which came out in East Berlin as well as in Vaduz/Liechtenstein; thus the volumes were available to readers in both the East and the West. The first volume, as mentioned, was authored by Helga Nussbaum herself, focusing on the nexus between the economic and political history of Imperial Germany. Volume two was written by Manfred Nussbaum and covered the crucial years between 1919 and 1933. As he did in his volumes on German colonial efforts in Africa, Manfred Nussbaum tried to avoid mere ideological phrases in favor of a more differentiated approach, describing and analyzing the complex and contradictory forces and interests that clashed as Germany's first bourgeois republic was overwhelmed by capitalist crisis, political paralysis, and the rise of fascism. While not free of ideological distortions, this volume constituted a valuable effort in combining economic, political, and social history in an overall non-dogmatic fashion.

The entire three-volume series *Wirtschaft und Staat in Deutschland: eine Wirtschaftsgeschichte des staatsmonopolistischen Kapitalismus in Deutschland vom Ende des 19 Jahrhunderts bis 1945* [*Economy and State in Germany: An Economic History of State Monopoly Capitalism in Germany from the End of the 19th Century to 1945*], written jointly by Helga and Manfred Nussbaum, as well as Lotte Zumpe and Dieter Baudis, was published in East Berlin by the Akademie Verlag and in Vaduz/Liechtenstein by Topos Verlag. The Liechtenstein edition appeared between 1978 and 1980 to generally positive reviews and illustrates the reach and impact of the work of those GDR scholars.

The Case of Jürgen Kuczynski

The most well-known GDR scholar, in the social sciences and humanities in general as well as in economic history more specifically, was, of course, Jürgen Kuczynski. He was also the best connected, especially with scholars in Western countries. Having lived for several years in the United States as a research student affiliated with the Brookings Institution and the American Federation of Labor, as well as in exile in Great Britain between 1936 and 1944, Kuczynski was familiar with the world outside the GDR and had many contacts there. He worked for the U.S. Army during World War II, held the rank of Lt. Colonel, and was associated with the United States Strategic Bombing Survey, working under the direction of John Kenneth Galbraith, with whom he maintained a lifelong contact. Kuczynski prided himself as a man of superlatives, with his family pedigree rooted in the radical Enlightenment, left-wing political activism, and academic distinction. Residing in a large thirteen-room villa in Berlin-Weissensee, Kuczynski was surrounded by his family library of over 70,000 books as well as 35,000 journals and magazines. His writing habits were as avid as his book collecting and reading inclinations. By the end of his life, Kuczynski had authored over 150 books of his own, augmented by approximately 4,000 papers, reviews, and essays.

The scope of Kuczynski's stature as a conduit for East-West contacts during the Cold War is illustrated by a conference organized in honor of the eightieth birthday of Marion Countess Dönhoff. The conference took place at a time when the Party's hold on the East German state was rapidly eroding. Scheduled for December 1–2, 1989, it was planned well in advance and, in addition to Kuczynski, featured an impressive list of important names from academia, journalism, and politics, as well as diplomatic officials. Chief among those invited to speak at the Dönhoff birthday symposium were the leading West German Social Democratic leader and former Federal Chancellor of West Germany, Willy Brandt; Peter Glotz and Egon Bahr, both senior members of the presiding council of the Social Democrats; Valentin Falin, director of the International Department at the Central Committee of the Communist Party of the Soviet Union; Kjell-Olof Feldt, Finance Minister of Sweden; Ralf Dahrendorf, formerly a member of parliament of the West German Liberal Party and, at the time of the symposium, the Warden of St. Antony's College in Oxford; and Henry Kissinger, the well-known and notorious former U.S. National Security Advisor and Secretary of State.

In addition to other politicians, state officials, and social movement leaders, such as Austrian Chancellor Franz Vranitzky and Adam Michnik of Poland, public intellectuals and polemicists like André Glucksmann of France, and high-profile and well-connected scholars, such as the historian Fritz Stern were on the circulated list of participants (Participants' List, 1989). Gerd Bucerius, publisher of *Die Zeit*, West Germany's leading liberal weekly and the institutional organizer of the symposium, asked Kuczynski to be co-chair of one of the three main panels. The other co-chair was Daniel Bell, whose diagnosis of an approaching *End of Ideology* stood in stark opposition to Kuczynski's call for a renewal of Leninism, with which he charmingly irked the other symposium participants (Bucerius, 1989). The fact that Kuczynski was invited to this symposium, given its role as a gathering place of the Western academic, political, and ideological elite, underscores his importance as a key figure in the intellectual exchanges between East and West during the Cold War. Arguably no other GDR scholar, let alone economic historian, had Kuczynski's connections, his intellectual reputation, and his overall stature.

Not all of Kuczynski's books were of equal quality. Yet, he provided an important antidote to the vulgarizations of Marxism in which both Party leaders and apparatchiks frequently engaged, attracting people to the GDR and Marxism who otherwise might have been repelled. An interesting case in point is Wolfgang Jacobeit (1921–2018), who earned his PhD in West Germany under Will Erich Peuckert in historical anthropology but later moved to the GDR and eventually worked with Kuczynski on the history of everyday life. As Kuczynski published his famous six-volume history of the everyday life of the German people, Wolfgang Jacobeit and his wife Sigrid eventually published their spin-off project; a three-volume illustrated social history of everyday life in Germany between 1900 and 1945.

Kuczynski is most well-known for his truly monumental multi-volume history of the working class under industrial capitalism. This massive project appeared in print between 1961 and 1972 and originally consisted of thirty-eight volumes. Later, two more volumes were added on child labor. The reviews were largely positive, especially in the East. But leading Western scholars, such as Jürgen Kocka from the influential Bielefeld school, also praised the work, recognizing its potential as well as its limitations. Kocka's verdict echoed what most well-informed and fair-minded Western colleagues concluded in their reviews as well. Concerning volumes one and two—the books reviewed by Kocka—"a better social history of the German working class in the age of the Industrial Revolution does not exist," and Kuczynski's effort stands, despite shortcomings and *Unwissenschaftlichkeit* [lack of rigorous scholarly standards] as a "pioneering effort" (Kocka, 1974, 471).

In addition to this monumental project on the condition of the working class under industrial capitalism, Kuczynski published two other large-scale projects: a five-volume *History of the Everyday Life of the German People* and a ten-volume *History of the Social Sciences and the Humanities*. Both projects characterized his style and methodology: immense industriousness, resourcefulness, and energy combined with a mix of politico-ideological dogmatism and subversion. Kuczynski was arguably one of the most erudite people imaginable, intimately familiar with the intellectual trends in economic history not just in the Soviet bloc, but in the English- and French-speaking worlds as well. Kuczynski read and wrote in both languages without effort. He introduced ideas, concepts, and analytical tools from both non-Marxist and independent Marxist scholars into his very Marxist works, derived in part from the

French Annales school as well as more culturally oriented historians within the orbit of the former Historian's Group of the British Communist Party. Yet, he did this unevenly. At times, Kuczynski was systematic, and at others, his approach was more superficial and bordered on name-dropping. He was on friendly terms with a host of highly influential American and British economists, ranging from John Kenneth Galbraith and John Maynard Keynes to Maurice Dobb. He corresponded with the key figures of the West German Bielefeld school and exchanged books and articles with many of the leading social scientists in the U.S., France, West Germany, and Great Britain.

In 1981 the Institute for Economic History published a two-volume handbook detailing what economic history was, how it fit into an overall system of knowledge and systematic inquiry, and the various specific historical processes and formations of economic development. After some introductory methodological reflections on periodization and the core conceptual structures of different economic epochs, the handbook elaborated on how the broad modes of production related to particular and regional historical variations. The authors tried to balance scholarly professionalism with the standard politico-ideological constraints that were expected in the GDR's official academic literature. Hence, in the second paragraph of the introduction, readers are reminded of the importance that "economic history has today, nationally as well as internationally, for the building of the developed socialist society, as well as for the global class struggle between socialism and capitalism, for the general instruction of working people as much as the education of scientists, for the further development of Marxism-Leninism as the theoretical foundation for the practical activities of the revolutionary party of the working class and in the struggle against bourgeois ideology in all its forms" (Institute for Economic History at the Academy of Science of the GDR, 1981, 11).

Commissioned by institute directors Wolfgang Jonas (1926–2010) and his successor Helga Nussbaum, nine colleagues from the institute were made responsible for the key sections of the two volumes: Berthold Puchert (1929–2016), Hartmut Harnisch (1934–), Thomas Kuczynski (1944–), Hans Radandt (1923–), Peter Musiolek, Siegfried Epperlein, Manfred Nussbaum, and Jörg Roesler (1940–). Hans Radandt, for example, was overall responsible for the section on historiography, sources, and working methodologies. However, the pages that focus on historiography itself, as well as its "pre-scientific" predecessors, including a brief overview of the leading international journals of economic history, were written by Jürgen Kuczynski. Given his stature as the legendary "nestor" of the GDR's Marxist scholars, the elder Kuczynski—not to be confused with his son Thomas, also an economic historian, professor at the Institute, and eventually its last director—seemed uniquely qualified to navigate this complex and ideologically loaded terrain. His international connections, seniority, reputation inside the GDR and outside, as well as his unique relationship with state and Party leader Erich Honecker gave Kuczynski leverage that few other East Germans possessed.

Kuczynski's historiographic essay sketches not only the contributions of Greek and Roman historians to economic history but also highlights Ibn Khaldun's work as well as other non-European historiographic traditions (J. Kuczynski, 1981, 208–210). Doing so illustrates, in however small ways, the efforts of GDR historians to free themselves from an exclusively Eurocentric gaze. The narrative nevertheless reverted to a more Eurocentric perspective, as Kuczynski moved from the ancient world and the Middle Ages into the European Renaissance and eventually the Enlightenment. He celebrated Hugo Grotius and Samuel Pufendorf as embodying the spirit of progressive early bourgeois rationality, just like the French Physiocrats. The most lavish praise, however, was reserved for Adam Smith. Kuczynski wrote enthusiastically: "[i]ndeed, I dare say that there are only two political economists who on such a grand scale succeeded in combining so skillfully theory with history: Adam Smith and Karl Marx" (J. Kuczynski, 1981, 218).

Kuczynski's segment on the German Historical School is also quite interesting, as it reflected not only the obligatory ideological norms required by the Party apparatus, but also his personal Marxist interpretations and positions as well as Kuczynski family intellectual traditions and experiences. Kuczynski introduced the German Historical School thus: "[t]here is a country that has succeeded in leading the world in economic history, at least for a while, and developed a unique historico-economic school based on a complete misunderstanding of how scholarly work should proceed…." This school, Kuczynski continued, has produced "remarkable results" thanks to strangely "mediocre scholars," which "illustrates how unevenly and in what crooked ways an academic discipline can develop" (ibid., 226).

Kuczynski's family background came into play when he referred to Lujo Brentano as "probably the best and most significant representative of the younger generation of the Historical School... Brentano was theoretically more interesting than Schmoller" (J. Kuczynski, 1981, 228). Readers familiar with the intellectual genealogies of economic history were thus reminded that Jürgen Kuczynski's father Robert René Kuczysnki earned his doctorate in 1897, under the supervision of none other than Lujo Brentano. Jürgen Kuczynski positioned himself, and by association the Institute for Economic History that he founded, as intimately rooted not only in Marxism-Leninism but also in intellectual and cultural traditions that went back into the nineteenth century and beyond. In doing so, he suggested, between the lines, that a genuine and innovative Marxism could not insulate itself without risking atrophy and fossilization. Kuczynski was cognizant of the basic reality that a Marxist tradition capable of producing new and relevant insights needed to seriously engage with non-Marxist currents, both past and present.

This was, of course, much easier said than done.

Conclusion

Kuczynski distinguished himself as an innovative as well as imaginative scholar and economic historian in East Germany. His scholarship, his personal contacts with colleagues around the globe and the Institute for Economic History that is so closely associated with his name were important conduits of contact and exchange between scholars from both sides of the Cold War divide. He, Hans Mottek, Günther Kohlmey, Helga and Manfred Nussbaum, and others worked toward overcoming the intellectual insularity of the GDR and helped its economic historians to participate in global exchanges.

Yet, whatever individual abilities, scholarly accomplishments, and international contacts and recognition they cultivated, their space for maneuvering was conditioned by the geopolitical context of the Cold War overall. The unique circumstances and predicaments of the German Democratic Republic limited how far East German scholars could go in developing and deepening ties to Western academic institutions and colleagues. The GDR's economic historians, who made up only a small fraction of their state's overall scholarly community, were bound by the tides of a volatile ideological and political landscape.

The GDR's political and diplomatic isolation, brought about chiefly by West Germany's relentless campaign of being the only legitimate German state in the 1950s and most of the 1960s, combined with the GDR's lack of freely convertible "hard" currency and made the latter's already controlling and paranoid Party and state leadership even more adverse to relaxing travel restrictions. Access to Western academic literature remained restricted for most East German scholars, as that material was seen by Party officials and administrators as both a considerable expense and an ideological risk. Thus, especially during the early decades of the GDR's existence, only a few well-established scholars with long records of loyalty to the Party were allowed to travel to conferences in the West, cultivate and maintain ties to colleagues and institutions there, or receive scholarly literature without interference. Eventually, as the GDR became recognized by more countries and even West Germany moved toward normalizing relations with its German neighbor, the GDR began gradually to expand the circle of scholars who were authorized to have outside connections. Thus, by the 1970s it was not merely Jürgen Kuczynski who represented the GDR's economic historians in the West at conferences and workshops but a growing cohort of younger scholars, such as Helga Nussbaum and Wolfgang Jonas. Yet, those who were "Reisekader" [allowed to travel], and hence, able to cross freely between East and West, remained a relatively small number of the GDR's scholarly community. That fact contributed to the sense of alienation and resentment that many East Germans felt toward the regime, even if they identified with the Marxist project.

References

Adolphi, W. (2022). *In Memoriam Gunther Kohlmey*. https://www.rosalux.de/fileadmin/rls_uploads/pdfs/Utopie_kreativ/112/112_Kohlmey.pdf. Accessed on 15 February 2022.

Agde, G. (1991). *Kahlschlag. Das 11. Plenum des ZK der SED 1965. Studien und Dokumente*. Aufbau-Taschenbuch-Verlag.

Bhambra, G. K., Gebrial, D., & Nisancioglu, K. (2018). *De-Colonizing the University*. Pluto Press.

Binghamton University. (2022). *Fernand Braudel Center*. https://www.binghamton.edu/fbc/about-fbc/intellectual-report.html. Accessed on 1 April 2022.

Blaschke, K. (1997). Als bürgerlicher Historiker am Rande der DDR: Erlebnisse, Beobachtungen und Überlegungenen eines Nonkonformisten. In K. H. Pohl(Ed.), *Historiker in der DDR*. Vandenhoeck & Ruprecht.

Bucerius, G. (1989, October 31). Hamburg. Letter to Jürgen Kuczynski, Berlin. Photo-Copy, Personal Archive Axel Fair-Schulz.

Fair-Schulz, A. (2006). Tons of Wasted Paper? Jürgen Kuczysnki and East German Historiography. In Q. E. Wang & F. Fillafer (Eds.), *The Many Faces of Clio: Cross-Cultural Approaches to Historiography Essays in Honor of Georg G. Iggers*. Berghahn.

Fair-Schulz, A. (2017). The Dissolution of the Institute for Economic History at the Academy of Science. In A. Fair-Schulz & M. Kessler (Eds.), *East German Historians Since Reunification: A Discipline Transformed* (pp. 155–164). SUNY Press.

Fischer, W., & Zschaler, F. (1998). Wirtschafts- und Sozialgeschichte. In J. Kocka and R. Mayntz (Eds.), *Wissenschaft und Wiedervereinigung: Disziplinen im Umbruch* (pp. 361–434). Akademie-Verlag.

Giroux, H. (2014). *Neo-Liberalism's War on Higher Education*. Haymarket Books.

Grieder, P. (2012). *The German Democratic Republic*. Bloomsbury Publishing.

Große Kracht, K. (2005). *Die zankende Zunft: Historische Kontroversen in Deutschland nach 1945*. Vandenhoeck & Ruprecht.

Hallgarten, G. W. F. (1967). Review of Unternehmer gegen Monopole: Über Struktur und Aktionen Antimonopolistischer Bürgerlicher Gruppen zu Beginn des 20. Jahrhunderts by Helga Nussbaum. *American Historical Review*, 73(2), 339–366.

Heitkamp, S. (2003). *Walter Markov: Ein DDR-Historiker zwischen Parteidoktrin und Profession*. Rosa-Luxemburg-Stiftung.

Heller, H. (2016). *The Capitalist University: The Transformation of Higher Education in the United States, 1945–2016*. Pluto Press.

Hesse, J. (2016). The Legacy of German Economic History: Archetypes and Global Diffusion. In F. Boldizzoni & P. Hudson (Eds.), *Routledge Handbook of Global Economic History*. Routledge.

Hoell, G. (1989). Mottek, Hans. In *Ökonomielexikon*. Dietz Verlag.

Iggers, G. G. (1983). *The German Conception of History: The National Tradition of Historical Thought from Herder to Present*. Wesleyan University Press.

Iggers, G. G. (1991). *Ein anderer historischer Blick: Beispiele ostdeutscher Sozialgeschichte, Published in English under the title: Marxist historiography in Transformation: East German social history in the 1980s*. Fischer Taschenbuch Verlag GmbH.

Iggers, G. G. (2017). Where Did Historical Studies in the German Democratic Republic Stand at the Eve of Unification? In A. Fair-Schulz & M. Kessler (Eds.), *East German Historians Since Reunification: A Discipline Transformed*. SUNY Press.

Institute for Economic History at the Academy of Science of the GDR. (1981). *Handbuch Wirtschaftsgeschichte*. VEB Deutscher Verlag der Wissenschaften.

Kocka, J. (1974). Review of J. Kuczynski, Darstellung der Lage der Arbeiter in Deutschland von 1789 bis 1849 and Darstellung der Lage der Arbeiter in

Deutschland von 1849–1870. In *Eine Auseinandersetzung mit dem Werk von Jürgen Kuczynski "Die Geschichte der Lage der Arbeiter unter dem Kapitalismus" Volumes 1–38*. Akademie-Verlag 1961–1972. (Special Print from the *Archiv für Sozialgeschichte*. Volume XIV, 1974.)

Kuczynski, J. (1981). Geschichtsschreibung der Wirtschaftsgeschichte. *Handbuch Wirtschaftsgeschichte*. (pp. 208–238). VEB Deutscher Verlag der Wissenschaften.

Kuczynski, J. (1992). *Ein Linientreuer Dissident: Memoiren 1945–1989*. Aufbau Taschenbuch Verlag.

Kuczysnki, J. (1994). *Ein hoffnungsloser Fall von Optimismus? Memoiren 1989–1994*. Aufbau Verlag.

Kuczynski, T. (2009). Rede zur Schliessung des Instituts für Wirtschaftsgeschichte der ehemaligen Akademie der Wissenschaften der DDR, gehalten am 7.10.1991 an der Akademie der Wissenschaften der DDR. *Jahrbuch für Wirtschaftsgeschichte*.

Loth, W. (1998). *Stalin's Unwanted Child: The Soviet Union, the German Question and the Founding of the GDR*. Palgrave Macmillan.

Miliband, R. (2009). *The State in Capitalist Society*. Merlin Press.

Participants' List. (1989, December 2). *Zeit-Symposium*, Hamburg. "End of Communism—What Now?" Photo-copy, Personal Archive of Axel Fair-Schulz.

Pohl, K. H. (1997). *Historiker in der DDR*. Vandenhoeck & Ruprecht.

Ringer, F. K. (1969). *The Decline of the German Mandarins: The German Academic Community, 1890–1933*. Harvard University Press.

Roesler, J. (2017). The Dissolution of East German Economic History at the Economic University in Berlin-Kalrshorst: A Typical Anschluss. In A. Fair-Schulz & M. Kessler (Eds.), *East German Historians Since Reunification: A Discipline Transformed*. SUNY Press.

Schneider, J. (1977). Marxistisch-leninistische Wirtschaftswissenschaften nach sowjetischem Modell an den Hochschulen der SBZ/DDR: Legitimation und Propaganda für Parteitage der SED In H.-J. Gerhard (Ed.), *Struktur und Dimension: Festschrift für Karl Heinrich Kaufhold*, Band 2: *Neunzehntes und Zwanzigstes Jahrhundert*. Franz Steiner Verlag.

Schulze, W. (1993). *Geschichtswissenschaft in Deutschland nach 1945* [*Historical Scholarship in Germany After 1945*]. Oldenbourg.

Steiner, A. (2004). *Von Plan zu Plan: Eine Wirtschaftsgeschichte der DDR* [*From Plan to Plan: An Economic History of the GDR*]. Deutsche Verlagsanstalt.

Strohe, H. G. (1996). Statistik in der DDR Wirtschaftsstudium, zwischen Ideologie und Wissenschaft. In *Statistische Diskussionsbeiträge*. (Vol. 3). Universität Potsdam.

Weber, W. (1984). *Priester der Klio: historisch-sozialwissenschaftliche Studien zur Herkunft und Karriere deutscher Historiker und zur Geschichte der Geschichtswissenschaft 1800–1970*. P. Lang.

Wolf, H., to Kuczynski, J. (1984, January 4). Berlin. Photocopy, Personal Archive of Axel Fair-Schulz.

CHAPTER 4

Czechoslovakia: Opening Doors to the West

Antonie Doležalová and Roman Holec

During the Cold War era, the Czechoslovak historiography of economic history neither represented a significant part of European historiography nor influenced its evolution. An integral part of economics in the late nineteenth century, the field of economic history in Czechoslovakia repositioned itself as a branch of history when Bedřich Mendl (1892–1940) established the first chair in economic history in the Department of History at the Faculty of Arts at Charles University in Prague in 1932. World War II had different effects on the slowly progressing discipline in then-divided Czechoslovakia. While the foundation of the Slovak Academy of Arts and Sciences in 1943 opened the door for historical writing defined as nationally Slovak, the Czech universities were closed down in the Protectorate of Bohemia and Moravia in 1939 and the Czech

A. Doležalová (✉)
Institute of History, Czech Academy of Sciences and Faculty of Social Sciences, Charles University, Prague, Czech Republic
e-mail: ad875@cam.ac.uk

R. Holec
Comenius University, Bratislava, Slovakia

A. Dolezalova and C. Albrecht (eds.), *Behind the Iron Curtain*, Palgrave Studies in Economic History,
https://doi.org/10.1007/978-3-031-31578-7_4

historical community gradually fell silent under the threat of repressions; Mendl was driven to suicide by racial persecution. His chair was never reopened.

After World War II, university departments, academies, museums, regional archives, and enterprises archives coexisted as workplaces of research in economic history in Czechoslovakia. However, there were only two departments of economic history: the Institute of Economic Geography and Economic History at the University of Economics in Bratislava established in 1951 and the Department of Economic History at the University of Economics in Prague established seven years later. At other workplaces, research in economic history was elaborated by individuals or small groups of specialists. The Silesian Institute that was created in Opava in 1953 focused on the industrial development of the Silesian region. Beginning in the 1960s, small groups of economic historians formed at the Institute of History of the Czechoslovak Academy of Sciences (hereafter the Academy) in Prague and Bratislava. At the same time, the Committee for the History of Enterprises was established as an accomplishment of the resolution of the Central Committee of Communist Party of Czechoslovakia (hereafter the Party). The Committee for Economic History subordinated to the Historical Society, which gathered the researchers in the field throughout the entire Czechoslovakia, and the Committee for Historical Demography associated with the Institute of History of the Academy in Prague came into existence in the 1966 and 1967, respectively; the Committee for Economic History was affiliated with the International Economic History Association (hereafter the IEHA). In 1972, an Institute of Social Sciences was established at the Slovak Academy in Košice, and its main focus was the socioeconomic development of eastern Slovakia and its minorities after 1945.

Even though economic history was a marginalized sub-discipline of history under inadequate institutional conditions, many historians addressed economic history in their research. Moreover, every historian in Czechoslovakia was confronted with economic history for the Marxist postulate of the primacy of the economic base over the superstructure. As a consequence, every research question had to take into consideration its economic aspects. As history had to serve the ideological and propagandistic needs of the regime, economic history was supposed to provide evidence for both the revolutionary tradition of the Czechoslovak people and the unavoidable trajectory of history toward communism (Doležalová & Holec, 2016a).

Although the Party accented need to elaborate the newest historical periods, contemporary archival sources were under the control of the Party apparatus and only accessible to a selected circle of the Party historians (Sommer, 2011). As researchers struggled to avoid strict Party censorship, medieval and early modern studies had a strong position in academic research at universities and at the Academy. Researchers concentrated on the economic development of feudalism, its crises and the transition from feudalism to capitalism; the various aspects of serfdom and manorialism; the economic development of towns and rural areas; the early stages of capitalism and the Industrial Revolution; and issues relating to class struggle. In the 1970s and 1980s, historians in enterprise archives were developing labor history and the history of enterprises, the socialist counterpart to the business history that was developing dynamically on the other side of the Iron Curtain. Despite the Marxist concept of the history, only a few historians took it seriously and employed a holistic Marxist approach in their research.

At first sight, Czechoslovak historiography during the Cold War era represents an example of a historiography that was relatively isolated and minimally exposed to external methodological influences. The ideologically pure methodology of Marxism-Leninism was losing its scientific nature and was unable to keep pace with the evolution of the world historiography of economic history. Yet, after close examination, we can find both personal and institutional ways of cooperation with foreign historiographies whether in the form of careful observation of the trajectories of development of historiography in other countries or through direct cooperation. Thus, this chapter challenges the narrative of Czechoslovak socialist scientists' incapability or disinclination to engage openly in international cooperation and discussions.

This study explores in what measure the state supported openness toward international cooperation and in what measure international cooperation represented an achievement of individual researchers. It also shows that Czechoslovak economic historians were influenced by or even cooperated fruitfully with economic historians from abroad. The chapter analyses the modes of transfer of ideas between Western and Czechoslovak historiographies, the ways of cooperation between Czechoslovak economic historians and their colleagues in socialist and Western countries, and the role Czechoslovak historians in exile played in this regard. Our analysis of the strategies historians used reveals how ambiguous these relationships were and shows that forty years of building socialism in Czechoslovakia were not changeless.

Watchmen and Protectors: Marxist-Leninist Purity and International Cooperation

After liberation in 1945, the majority of political parties associated with the governing National Front in Czechoslovakia accepted socialist ideas and included planning, nationalization, and the liquidation of quislings, traitors, and also German and Hungarian business elites in their programs. The postwar economic structural changes were realized through the nationalization of key industries and the Two-Year Plan for state-owned enterprises; in the public sphere these steps were considered a unique Czechoslovak path to democratic socialism. Three years later, when the communists completed their journey to political power in February 1948, Czechoslovakia departed from that path and embarked on the way to socialism of the Stalinist type. A total nationalization of all industrial sectors and services, implementation of an administrative-directive system of planning of the entire economy, violent collectivization of agriculture, devastating monetary reform, and oppression of churches followed. In 1945, the economically most advanced country in Central and Eastern Europe started to lose its economic dynamism and efficiency and to balance at the edge of decline to the level of developing countries. Two attempts to change the trajectory by economic transformation failed in 1957 and 1968. In the official narrative, the entire Western civilization had become imperialistic and capitalism was a decaying system. The Party became the only leader of the nation and its members penetrated all administrative and political institutions. Marxism-Leninism, defined in conformity with the official Soviet narrative as the theory of the socialist revolution and the socialist economy, became the only scientific methodology.

However, the situation in science was not always equally oppressive. For our analysis of the international scientific cooperation of Czechoslovak economic historians, we divide the socialist years in Czechoslovakia into three periods. To explain the state of historiography, we have to start, however, with short description of the three years of post-war reconstruction between 1945 and 1948, when historians struggled to return to prewar scientific practice. There were no obstacles to international cooperation or the free exchange of ideas; beside resumption of *Český časopis historický* [*Czech Historical Review*], which had been established in 1895 and banned during the Protectorate, also *Sborník pro hospodářské a sociální dějiny* [*Review of Economic and Social History*] was published by the Central Committee of Unions in the years 1946 and 1947. Nevertheless, historiography was paralyzed by a clash between non-communist and pro-communist historians about the new narrative of the national history. The pro-communist narrative combining ideological elements with selected elements of the national movement of the nineteenth century and using key words such as "logic of the progress" or "truth of the history" prevailed. The battlefield par excellence was the Hussite movement of the fifteenth century as it demonstrated an excellent example of both the fight with (German) enemies and the roots of the revolutionary attitude of the Czech nation (Doležal, 2022). From that point on, scientists were referred as the "theoretical front."

At the same time, there were only a few Marxist historians in Czechoslovakia: members of Historická skupina [Historical Group] that had been established in the 1930s. Its political background, however, was of a social democratic rather than a communist nature. The methodological perspective of its members inclined to the Marxist interpretation of the French Annales school rather than to Marxism-Leninism of the Soviet style. In general, knowledge of the ideas of Marx, Engels, and Lenin was only modest in the Czechoslovak environment and was largely subordinated to the Party's popularization of the Russian Revolution and its critique of the bourgeois character of interwar Czechoslovakia. In this narrative, Marxism was transformed into Marxism-Leninism, applying Marx's theory to specific conditions in Russia. Thus, translations of Marx's books were accompanied by "re-narrating" Marx's teaching to make Lenin's contribution apparent.

The first period (1948–56) of the communist era in Czechoslovakia started with the communist takeover followed by a wave of political purges and ended with the critique of the cult of Stalin's personality. The period was characterized by the Stalinization of the organization of science. The interpretation of history was gradually subordinated to political and ideological assignments of the Party according to which any scientific production was supposed to be an instrument of mass education and a gun in the hands of the working class. Marxism-Leninism was transformed into Stalinism, in which Marx's original teachings mutated into a vulgar and dogmatic form. From that time, for the large majority of historians, the application of historical materialism started and ended with a direct quotation from Marx, Engels, and Lenin in the first paragraph of their works.

The strict censorship implemented by Soviet experts guarded compliance with Marxism-Leninism. Over two hundred non-communist newspapers and scientific journals were liquidated, including *Czech Historical Review*. The import of foreign scientific journals was restricted. However, economic history did not suffer from these restrictions: the main journals in the field, *Annales d'histoire économique et sociale, Vierteljahrschrift für Sozial- und Wirtschaftsgeschichte,* the *Economic History Review,* and the *Journal of Economic History,* were still available at least in the Historical Library at the Faculty of Arts at Charles University and in the University Library in Prague. In 1953, new journals in historical sciences, *Československý časopis historický* [*Czechoslovak Historical Review*] (hereafter the *Review*) in Prague and *Historický časopis* [*Historical Journal*] in Bratislava were established. Despite their dedication to Marxism-Leninism, they informed readers not only about the achievements of Marxist-Leninist historiography but also about the newest developments in world historiography. *Historické štúdie* [*Historical studies*], published as a yearbook in Bratislava beginning in 1955, was an even more important journal for the field of economic history.

New institutions were established to supervise scientific research. Besides the Party's local committees, the unions operated in all workplaces and membership in them was compulsory. A specific role in popularizing the achievements of socialist science was played by the Socialist Academy, which was renamed the Czechoslovak Society for Dissemination of Political and Scientific Information in 1952 (in public, it was ironically called the long-name-society) and returned to its original name in 1965 (Olšáková, 2014). In 1950, the Institute for the History of the Czechoslovak Communist Party was founded for the purpose of creating an official narrative of the history of both the Party and the class struggle (Sommer, 2011). The Academy of Sciences was founded in 1952 as a demonstration of the transformation of Czechoslovak science according to the Soviet model, with its distinction between education and research. Its Institute of History in Prague was divided into departments of feudalism (till 1848), capitalism (till 1918), and modern history; two-thirds of its 21 original employees were Party members.

The crucial institution for the implementation of Marxism-Leninism into science was the College of Political and Economic Sciences, which was established as a "pioneer Marxist school" and was managed directly by the Party's Department of Propaganda (Doležalová, 2018). In the spirit of positive discrimination in favor of workers and peasants, its students often had incomplete secondary education. Despite its short seven-year history, the College's graduates and its more than two hundred educational texts implemented Marxism-Leninism throughout the entire educational and scientific environment as the new official "scientific" doctrine. Some of the College's graduates and young teachers obtained positions in history departments; the majority of them flooded the departments of Marxism-Leninism that were established at all universities in Czechoslovakia. From that time, only rarely did non-Party members obtain a leading role in the organization of research and with the exception of a short interval in the mid-1960s only Party members could teach at universities. By contrast, prominent Party members often were named university professors without standard academic procedures or credentials.

To avoid confrontation with the nomenklatura, the majority of researchers moved from modern history toward older historical periods where they expected—and found—less oppressive working conditions. In all areas of research, however, international cooperation was almost impossible until Stalin's death in 1953. The door was opened through mutual participation at national historical congresses of socialist countries orchestrated by the Soviet Union; for the first time historians met in Budapest at the Congress of Hungarian Historians in 1953.

The second phase (1956–68) was characterized by easing the Stalinist grasp and searching for a Czechoslovak socialism with a human face; it ended with the termination of the Prague Spring. Historiography, however, witnessed a short period of criticism of Western novelties at the end of the 1950s when mainly non-communist historians were marked as revisionists and bourgeois opportunists. As a consequence, historians used such terms as comparative analysis, structuralism, and other "bourgeois" innovations in historiography cautiously until 1989 with the exception of the short period of the Prague Spring during which historians treated Marxism as a relevant methodological tool for the first time.

Encounters with international developments in historiography were intensified as libraries increased their purchases of foreign books and journals not only from socialist countries; library holdings in economic history were broadened by *Jahrbuch für Wirtschaftsgeschichte, Scandinavian Economic History Review, Business Archives and History, Journal of the Economic and Social History of the Orient, Indian Economics and Social*

History Review and *Economy and History*, which were available at least in the University Library in Prague.

The options for wider cooperation with Western colleagues also flourished. Economic historians traveled abroad for conferences, seminars, and study trips; first to the other socialist countries and later to Western countries as well. The procedure for gaining financial support or approval to travel abroad remained, however, under strict Party control. Historians used the opportunities to travel also on individual bases, investing their private money or utilizing their foreign relations or foreign scholarships. From the mid-1960s, the courage of Czechoslovak historians increased, albeit slowly, to enter international debates, deliver lectures at foreign universities, participate in international projects in the West, publish the outcomes of their research abroad, and present their papers at international conferences.

No matter how minimal the obstacles to traveling were, ideological supervisors did not stop zooming in on activities of exile historians who had been officially expelled from the community of Czechoslovak historians at the end of the 1940s. Even in the most liberal 1960s, any scientist who was selected to travel to the West had to sign a declaration avoiding exile Czechoslovaks, often his or her former colleagues. After their return, scientists had to report to the state police about each person with whom they met abroad. (Doležalová, 2021).

The third period of normalization (1969–89) tightened the screw again. The period was opened with a new wave of purges at academic workplaces; up to 145 historians had to leave their positions in academia (Prečan, 1994) as they either participated in the Prague Spring movement or refused to accept the interpretation of Soviet occupation as a liberation. They were mostly transferred from academia to the working-class sphere; the luckiest of them found positions in museums and enterprise archives that, as a consequence, started to play an important role in economic history research (Aston & Hilton, 1974). Some expelled scientists chose to emigrate. Several of those who stayed in academia became zealous "normalizers," adjusting historiography to Marxist-Leninist purity and liquidating any Western elements in them.

Self-censorship obstructed open discussion. The Party censorship, which was in the hands of non-historians, supervised the content of academic articles and books. Even the themes of masters and doctoral theses were the subject of ideological supervision. Ideologically incorrect books were withdrawn from libraries; however, journals and books in economic history remained available and library portfolios were extended by the *Journal of European Economic History, African Economic History, Business and Economic History, Irish Economic and Social History, Histoire, Économie et Société, Scottish Economic & Social History, Essays in Economic and Business History, Journal of European Economic History,* or *Explorations in Economic History*. In 1978, the Academy started to publish the journal *Hospodářské dějiny* [*Economic History*]; in the 1980s, Czechoslovak contributions at the WEHC were published in issues of *Economic History*.

Traveling abroad came under careful scrutiny again with many peculiar parameters. The breakpoints were already visible at the Congress of Historical Sciences and the World Economic History Congress held in Moscow and Leningrad, respectively, in 1970. The chair of the economic history delegation in Leningrad, Arnošt Klíma (1916–2000), was dismissed by the Czechoslovak Minister of Education when he was already in the USSR; only immediate intervention by the congress president, V.A. Vinogradov, allowed him to stay. Despite the hope for détente and extending cooperation after the Helsinki accords (Berg, 2015), historians in Czechoslovakia were overwhelmed by the rising pressure of self-censorship after the *Charta 77* was issued in 1977. The pressure of the regime increased again. Not only were historians asked to express their rejection of the *Charta 77*; they were also forced to even stronger equivocation and compromises when formulating the results of their research. It was not until the second half of the 1980s that ideological tightening slowly started to ease, thanks to Soviet perestroika.

During the entire communist period, international cooperation was an ambiguous activity for Czechoslovak scientists. To continue the parable of the Iron Curtain as a barricade, described at the beginning of the book, international congresses were battlefields in the fight between the only proper Marxist historiography and its bourgeois enemies, and meetings of historians from socialist countries were preparation grounds for loading the weapons for the battles. Historical journals were more or less open windows to the West, which provided fresh air of inspiration for keeping with the progress in the field in the non-communist world.

Czechoslovak historians participated in both congresses of historical sciences and congresses of economic history. At these events, Czechoslovakia was represented officially by between two (in Paris in 1950) and 251 historians (in Vienna in 1965). Five historians participated in Rome in 1955; three of them included economic-historical aspects in their papers.

In Stockholm in 1960, the Czechoslovak delegation of 31 members was the third largest from the Soviet bloc, trailing only the fifty members of the Soviet and the East Germany delegations (Kladiva, 1961, 151). However, at the meeting of economic historians in the framework of the congress, only Arnošt Klíma delivered a paper. The modest Czechoslovak representation in the meeting and no Czechoslovak participation in the meeting's preparation (unlike Polish historians) provoked strong Party criticism toward the state of economic history in Czechoslovakia and led to the Party's declaration requiring to establish the Committee for Economic History (Kladiva, 1961). In Aix-en-Provence in 1962 Czechoslovak historiography was represented by Jaroslav Purš (1922–1997) and Arnošt Klíma; Václav Husa (1906–1965) and Josef Petráň (1930–2017) sent their papers to be read (Purš, 1963). The large delegation traveled to Moscow and Leningrad. Two years after the invasion of Czechoslovakia, the Soviets strove to achieve numerical superiority of historians from Socialist countries to avoid unpleasant discussion; 22 Czechs and Slovaks were members of the official delegation and 137 other nominated historians participated in the congress in Moscow (Erdmann, 2005, 249–251; Jiroušek, 2006, 75; Polišenský, 2001, 262).

Even after the establishment of economic history congresses in the 1960s, Czechoslovak economic historians participated in congresses of historical sciences as well. In some measure, it was caused by the fact that themes for congresses of historical sciences included questions and topics that are considered the domain of economic history, such as the crisis of feudalism, social movements, working conditions, or the Industrial Revolution. But the fact that since 1968 economic historian Jaroslav Purš was the chair of the Czechoslovak delegations for any congress and an editor of the volumes of Czechoslovak contributions at congresses published in English had a decisive influence as well. The decision of who would represent Czechoslovakia at the world congresses was always a political question in Czechoslovakia.

Strict ideological criteria played the main role in the selection of official delegation members, while the expertise of historians played only a limited role. As a consequence, to secure delegates who were also experts in speaking foreign languages was not always easy. For instance, world-known historian Josef Polišenský (1915–2001) was not allowed to travel to Paris in 1950 despite the invitation he received from the organizers (Polišenský, 2001, 167–168). He was replaced by two loyal Party members confirmed by the Central Committee of the Party. Despite

Klíma's position and prestige in the IEHA (Berg, 2015), he mostly traveled for congresses on his own and not as a member of the official delegation. Only in the Prague Spring period did Klíma take the lead in organizing Czechoslovak participation at the WEHC.

From 1953, the partly confidential meetings of the highest representatives of historiography from socialist countries under the baton of the Soviets that were organized before international congresses started to influence the composition of national delegations. As already the first meeting in Budapest ordered the creation of bilateral historical committees, Czechoslovak historians established committees with their colleagues from the GDR, Poland, Hungary, the USSR, and Yugoslavia in the following years, along with a few multilateral committees. In the second half of the 1960s, joint meetings with Western historians occasionally took place, including a rare colloquium of Czechoslovak, Austrian, and West German historians in 1967. However, the functioning of committees depended on who was named as chair; some committees were only formal bodies while others provided a collaborative atmosphere. For instance, the Czechoslovak-Polish Committee under the lead of Polish medieval historian Roman Heck (1924–1979) organized workshops within regular meetings of the committee. Thanks to the close relations of Polish historians with their western colleagues, Western historians also participated at the workshops when they took place in Poland. In the mid-1970s, the workshops focused on historical consciousness, a topic that was in the very early stage of research at that time. When Czechoslovak Polonist František Hejl (1920–2016) took over the chair in the early 1980s, he changed the focus to the newest history and the meetings became ideologized (Hroch, 2023).

From 1957, the Czechoslovak Historical Society (hereafter the Society), which had been established that year as a part of the Academy, organized Czechoslovak participation at the world congresses, including proposals for congress panels. The Society collected suggestions for panel proposals from all academic workplaces in the country. However, until 1989, only departments chairs and the Party chairs at workplaces were involved in the process. Before the congresses, all papers to be presented were double-checked to ensure that they demonstrated the proper expertise and ideological maturity (Jiroušek, 2006, 61–62). As the number of participants grew, ideological supervisors were appointed to join a delegation; they were usually members of the Ideological Department of the Party who did not speak a foreign language and were not experts in the field. However, they attended any panel with a Czechoslovak paper and checked that historians read exactly the paragraphs approved before the congress. They indeed recognized if the names of Marx or Engels or some paragraphs from the written text were missing or even replaced by different paragraphs that had not been approved.

Another important aspect that has to be taken into consideration when evaluating the Czechoslovak delegations at congresses is the availability of foreign exchange. Foreign exchange was in short supply in communist Czechoslovakia and was allocated within a five-year plan for scientific research at individual workplaces, which financed the international trips of their employees. The trips for those who were not members of the official delegation were organized through the state-owned travel agency Čedok. A curious situation happened during 1965 historical sciences congress in Vienna. The number of participants was not strictly scrutinized that time due to the ongoing process of the Prague Spring. Beside an official delegation of 27 historians and 14 others who traveled on their own, Čedok organized the travel and accommodation for 210 other delegates from various Czechoslovak workplaces. To minimize the price, Čedok rented a special cruise ship on which delegates came to Vienna from Bratislava and lived as well. Thanks to that, the delegates did not need foreign exchange, since also food could be paid for with Czechoslovak crowns on board (Jiroušek, 2006, 67).

Due to the multiple obstacles to travel for most Czechoslovak economic historians, the only sustainable means to partially breach the insurmountable Iron Curtain remained domestic historical journals. From their pages, historians could learn about foreign literature: Hungarian, Polish, and German books reviews dominated in *Historical Journal* issued in Bratislava, while the *Review* covered the full range of international publications. Both journals also published reports from foreign conferences. In any issue of the *Review*, up to one-fifth of the pages was dedicated to reviews and reports. This meant up to thirty pages of reviews and five pages of reports with a significant share devoted to economic history. Already in its first volume issued in 1953, the *Review* published an article on the newest developments in French and British economic history, focusing on history of the working class. From the very beginning, authors always used the rhetoric of criticism toward Western historiography in the titles of their reviews and reports ("critical review," "relic bourgeois principles," and so on) and the proper words in the texts ("ideologically unexceptionable," "democratic principles of the

Soviet science," and so on) under the cover of which hitherto tabooed themes and figures (including Sudeten-German and Czechoslovak exile historiography), unknown historiographical innovations and the work of distinguished Western economic historians broadened the horizons of Czechoslovak historians; reviews of exile historiography were possible only with a narrative of criticism toward renegades. Besides that, short descriptive overviews of historiographies that were at the periphery of the world mainstream that time, such as Spanish, Asian, or Latin American historiography, enriched the Czechoslovak environment. Even though the leading foreign journals in the field were available in the main libraries, the *Review* provided a regular summary of their volumes as well.

Through these articles, even those who did not speak the foreign language could learn about the state of the art in the field. Since almost every Czechoslovak historian spoke German at that time, this applied particularly to English-language publications. The only significant and visible period of decline in informing openly, yet ideologically, about the progress of the historical sciences in the West were the years 1977–78 that followed the declaration of *Charta 77*. Both volumes of the *Review* were full of articles on the sixtieth anniversary of the Russian Revolution and on the progress of Soviet and socialist historiography; moreover, almost exclusively Soviet books were reviewed in these years.

We cannot point out that everyone could work with books reviewed by *Review*; especially historians at universities outside of Prague and in local enterprise archives could face significant obstacles in accessing foreign publications. Even so, the quality of reviewing in the *Review* is a strong indicator that the broadest community of historians could have been familiar with the progress in their field because every historian read the *Review* as it was the only journal in the country that covered all branches and time periods of history. In addition, the faculties libraries provided a system of interlibrary and international loans, as individual workplaces had no foreign currencies to purchase foreign publications directly. The system of loans worked for free and quickly as every university department has an employee responsible for such services. To obtain a book from any European library took approximately one month. However, only a small number of historians used the service.

Varieties of East–West Cooperation in Three Case Studies

The above observations about Czechoslovak historiography reveal that while traveling abroad was a highly political issue connected with ideological and bureaucratic obstacles, economic historians were regularly updated with new developments in world historiography through the main historical journals. Thus, the question arises in what measure individual economic historians were open to implement foreign developments into their research and to embark on international cooperation with their colleagues either in other socialist countries or in the West. This subchapter explores the strategies of a few Czechoslovak historians who accepted the challenge to cooperate with their colleagues behind the Iron Curtain and analyses the means they utilized to transfer ideas.

Based on archival research, biographies, memoirs, and interviews with scientists active during the communist era we zoom in on (1) Alice Teichova, who emigrated after the Soviet invasion of 1968 and thus experienced East–West cooperation from both sides of the Iron Curtain; (2) Miroslav Hroch, who faced obstacles to international cooperation from the east side of the Iron Curtain and yet achieved an international recognition; and (3) general patterns of the transfer of ideas among historians who were exploring the process of the Industrial Revolution.

Alice Teichova (1920–2015): As If There Were No Borders

From the first major academic publications written in Czechoslovakia through hundreds of articles and books written in exile, the scientific work of Alice Teichova became a classic in the historiography of economic history. From both the eastern and the western side of the Iron Curtain, she collaborated with many distinguished figures from universities around the world. Her life was full of fateful twists and existential choices at the crossroads of twentieth-century history, and her story reveals the multifaceted and painful experience of exile. It was affected by the Second World War, the Holocaust, communist ideology, and the Cold War. Her only stable anchor was her husband, Mikuláš Teich.

Alice Teichova was born in Vienna. After Germany's Anschluss with Austria in 1938, she emigrated to England where she studied economics at the University of Leeds. In 1944, she married a Czechoslovak exile, Mikuláš Teich (1918–2018), who studied chemistry at universities in

Exeter, Nottingham, and Leeds. Not only love but also their leftist conviction connected them for the rest of their lives. In Exeter and Leeds, they participated in the communist youth movement and became members of the Czechoslovak Communist Party in exile.

After the war, Alice had no hesitation about following her husband to Czechoslovakia with enthusiastic expectations about building communism. She learned the new Slavic language, began teaching general history at Charles University's Faculty of Education in 1950 and continued her studies; she earned her master's degree (PhDr) in history from Charles University in 1952. In the same year, Mikuláš was expelled from the Party, being the victim of a politically motivated purge. He was accused of cooperation with Jewish imperialists in England; that accusation referred to the Montefiore family that had financially supported his university studies during the war. After months of fright, Mikuláš found a position as a researcher in the newly established Academy in 1952—in the institution established to build the Marxist-Leninist interpretation of history! (Doležalová, 2021). Alice was awarded the title of Candidate of Sciences (CSc, the socialist counterpart of PhD degree) in 1960 and in 1964 she completed her habilitation in economic history (Doležalová, 2016).

In August 1968, just after the invasion of Warsaw Pact troops, the couple slipped through the Iron Curtain at the very last minute before it shut. They left Czechoslovakia for a long-planned study trip to the United States, as Mikuláš had obtained the Academy's scholarship for a one-year visit to leading American universities Harvard, Yale, Wisconsin, and Berkley. In the turbulent time after the Soviet occupation, they delivered lectures, gave interviews, and visited exile organizations. In six months, their lives were changed by the invitation Mikuláš received from Cambridge. After their second emigration, in 1970, England became their home once again, this time for good.

Alice began teaching political and economic history at the University of Cambridge. She spent a year as a Visiting Fellow at University College (now Wolfson College) and three years as a Visiting Bye-Fellow at Girton College, where she was named an Honorary Fellow in 1989. In 1970, she was appointed Senior Associate Member at St Antony's College in Oxford for a year to participate in the research project on the economic history of Eastern Europe led by Michael Kaser (Kaser & Radice, 1985–86). In 1972, she decided to take the post of lecturer in economic history at the University of East Anglia in Norwich. In 1973, she advanced to the position of reader, and in 1975 she became the university's first female professor. In 1974, Cambridge University Press published her groundbreaking book, *An Economic Background to Munich*. Over the course of the 1970s and 1980s, she turned the University into a center of European economic history. She continued to pursue her research interests: Czechoslovakia and Central Europe during the interwar period and its industry, banking, capital flows, foreign trade, cartels, multinational corporations, currency issues, the state of the economy, and economic thought in the region. In most of these subject areas, her books and articles were pioneering and became the foundation for other studies (Doležalová, 2016).

The Teichs' story was and at the same time was not extraordinary: the Nazi occupation forced into exile a number of Jewish students and young people from Central Europe while the Soviet occupation forced into exile many Czechoslovak intellectuals. At the same time, the Teichs' story demonstrates the bizarreness of the communist regime and, as such, it differed from others in many ways. First of all, during the strong wave of immigration to the West from Czechoslovakia after the communist coup d'état, the Teichs decided to relocate to Prague to build a better world. What awaited Mikuláš in Czechoslovakia, however, was his expulsion from the Party and an eleven-year fight for rehabilitation. Alice remained a Party member for the entire time she lived in Czechoslovakia; moreover, she became head of the Department of History at the Faculty of Education in 1959 and a chair of the communist organization at the faculty in 1967. Her husband—the traitor—was not an obstacle to obtain such a high position in the nomenklatura.

Second, as both Teichs graduated in England during the war, their scientific attitude gained international impulses, including Marxism. Therefore, not only two devoted communists but also two well-educated young Marxists came back to postwar Czechoslovakia. A key teacher for Alice was the Marxist economist, H. D. Dickinson, who was her lecturer in Leeds. In the following years, the Teichs kept up with the newest developments in their fields in the West. They emphasized that even in the most politically oppressed 1950s Western literature was available in the University Library in Prague. Full of his typical enthusiasm, Mikuláš tried to arrange the import of books and journals to Czechoslovakia as the correspondence among Mikuláš, Joseph Needham, and a few scientific institutions in England reveals; he offered barter exchange as the Institute of History had no foreign exchange to purchase them. The offer was rejected by the British side because Czechoslovak books and journals were written in Czech (Doležalová, 2021). The Teichs also published numerous reviews and reports about newly published books in the West. Despite the prevalent view that the textbooks of that time either fulfilled the Party assignments or were translated from Soviet originals, there were no objections when Alice offered to translate books from English and collected a group of young historians to write the original Czechoslovak textbooks on economic history in the 1950s and 1960s (Teichova, 1960, 1968).

Third, the Teichs never stopped developing their personal scientific relations with Western scientists. As they repeatedly emphasized, Czechoslovak historians had options to cooperate with Western historians, but they were not interested in doing so. The Teichs, by contrast, used any opportunity to meet foreign colleagues. They recollected the visits of Sidney Pollard, Hans Raupach, Georg Eckert, Georg Iggers, and Wilhelm Treue in Prague; Treue participated in the Committee for History of Enterprises, and the Central Council of Czechoslovak Unions financed his stays in Prague despite the fact that Treue was that time an archivist in the Salomon Oppenheimer Bank (Doležalová, 2021).

Even in the years of Mikuláš's expulsion from the Party, both of them visited the United Kingdom repeatedly. The rule for traveling was clear: the Teichs were never allowed to travel jointly. When they traveled jointly (for the first time for the WEHC in Munich in 1965), their children had to stay at home. Alice with their children visited her brother, who was a professor at Edinburgh University, every summer. During these trips, she always visited British academicians, even though she did not inform the Czechoslovak secret police about it back home. In the 1960s, Alice negotiated with British Marxist historians, members of the Communist Party of Great Britain, to organize a joint panel for the congress in Vienna in 1965 (Jiroušek, 2006, 65). In the 1950s, when Mikuláš was researching Charles Darwin, Alice visited Darwin's granddaughter Lady Barlow to ask her for further information about Darwin's biography. Lady Barlow even accepted an invitation to come to Prague and deliver a lecture at the Academy (Doležalová, 2021).

Throughout the 1950s and 1960s, the Teichs traveled to other socialist countries for seminars, conferences, and research visits; already in 1957 Alice started an exchange program for students and professors with the university in Potsdam. In 1965 and 1968, she participated officially at the WEHC. Coincidentally, Alice presented the paper on foreign investments in Czechoslovakia that she later developed into *An Economic Background to Munich* at the congress held in Munich. The paper became her best foreign investment in her relations with western colleagues; two years later, Alice sought out Peter Mathias in Cambridge, who remembered her paper very well and invited her for the WEHC in Bloomington in 1968; the organizers covered her travel expenditures.

Fourth, the Teichs also kept an active correspondence with their foreign colleagues. Without doubt, these letters were read and checked by the secret police, who created the file on Mikuláš in 1952 before his expulsion from the Party. However, as contacts with Marxists and communists predominated in their correspondence and their letters had a strong Marxist content, the Teichs were soon considered rock-like Marxists. After the family's emigration, the secret police informed their employers that the Teichs were not coming back to Czechoslovakia when Mikuláš became a visiting fellow at King's College and Alice a Bye-Fellow at Girton College. The secret police based its conclusion on the fact that the couple did not apply for the new passports as they were required to do. The secret police created a files on the entire family and kept them until the 1980s.[1]

The state of the Teichs' cooperation with colleagues behind the Iron Curtain changed completely after their emigration in 1970. For a few years, Alice still considered returning to Czechoslovakia. With varying degrees of success, they struggled to keep in touch with former Czechoslovak colleagues and friends. Mikuláš even gathered Czechoslovaks for the first meeting of the International Committee for the History of Technology at Pont-a-Mousson in France in the 1970; while Jaroslav Purš accepted his invitation, Milan Myška (1933–2016) was already not allowed to travel abroad as he was expelled from academia after the Soviet invasion in 1968.

As personal contacts were only limited, the primary way to maintain relations was through correspondence (Doležalová & Holec, 2016b). In the first years, the Teichs corresponded with many of their former colleagues; in the course of time the circle was narrowed to a few of them:

[1] Archiv bezpečnostních složek, Praha, Archivní protokol operativních svazků centrálních útvarů MV a KS MV Praha 585237-61436; Archivní protokol vyšetřovacích spisů 14581–16819.

Luboš Nový (1929), Josef Smolka (1929–2020), Pavla Horská (1927–2021), Jaroslav Folta (1933–2011), Arnošt Klíma, Jaroslav Pátek (1934–2003), Milan Myška, Július Alberty (1925–2019), and Ákoš Paulinyi (1929–2021), who emigrated to Germany. Through the correspondence, the Teichs were not only informed about dramatic developments in Czechoslovak academia and society; they also continued discussions on the research topics. Interestingly, also Klíma, a historian internationally renowned but forced to the periphery at home asked for feedback on his writing on the Industrial Revolution (Klíma, 1986a, 1986b). Mikuláš answered in detail and Klíma implemented his recommendations.[2]

Throughout twenty years, the Teichs continued to exchange books and journals with their Czechoslovak colleagues. Even more, the Teichs subscribed to *Past & Present* for Klíma; the journal was delivered to his private address.[3] Likewise, Czechoslovaks used to send the newest Czechoslovak books to the Teichs. Mikuláš particularly asked for Czechoslovak books and articles on brewing and the scientific revolution published after 1968, topics that he elaborated over a long period. Alice asked for books and articles on Czech economic thought when she wrote a paper for the Fondazione Luigi Einaudi conference[4] and when she was editing the special issue of *The History of European Ideas*.[5]

The Teichs invited their former colleagues to participate in research and publishing projects. Only a few of them accepted the invitation: Folta, Smolka, Klíma, and Pauliny (Teich & Porter, 1986, 1992). Alice translated texts by Klíma and Myška that were then published by *Economic History Journal* and *Past & Present* (Klíma, 1979; Myška, 1979). As she recollected, it took three years to complete the manuscripts. Alice also helped Czech historian in German exile, Karel Kaplan(1928–2023),

[2] Girton College Archive, GCPP Teichova 2/6/16. Academic correspondence of Alice Teichova and Mikuláš Teich: File 16. Klíma to the Teichs, March 6, 1984; Milkuláš Teich to Klíma March 25, 1984.

[3] Girton College Archive, GCPP Teichova 2/6/18. Academic correspondence of Alice Teichova and Mikuláš Teich: File 8. M. Teich to Past&Present, October 6, 1976.

[4] Girton College Archive, GCPP Teichova 4/6/5. Papers on International Finance—Interwar Period.

[5] Girton College Archive, GCPP Teichova 3/6. History of Economic Ideas.

find a publisher for his book on the crimes of communism and wrote its introduction[6] (Kaplan, 1986).

In their articles and books, the Teichs used any opportunity to mention their former Czechoslovak colleagues' work. Alice often quoted economists connected with the Prague Spring movement who were banned by government in Czechoslovakia that time. By contrast, the Teichs' books were not quoted, let alone reviewed in Czechoslovakia during the 1970s and 1980s, including Alice's *Munich,* a few copies of which were in the University Library and libraries of Institutes of History in Prague and Bratislava. In 1976, Milan Myška wrote the Teichs that he learned about the book from his Polish colleagues.[7] No reviews on or references to *Munich* appeared in Czechoslovakia until 1989. The same treatment also applied to Alice's and Mikuláš's further books including *International Business and Central Europe 1918–1939*, which the Institute of History library purchased in the year it was published (Doležalová & Holec, 2016a).

After their emigration, the Teichs were ignored in Czechoslovakia. The reason why is not, however, one-sided. As the secret police was well informed about the Teichs' participation in international congresses, the Czechoslovak historians who took part in the same events were asked by the secret police immediately after their return home whether they met them. Marxist émigré Teichs were persona non grata for Czechoslovak communists. It took courage to be associated with the name Teich. This sheds new light on the behavior of Czechoslovak historians at the congress in Edinburgh in the story told at the beginning of the book. The only person who openly talked to the Teichs was Jana Engelová, former student of Alice. As she recollected, she told the secret police it would be impolite to evade her former professor.[8] In the 1980s, when Helga Nussbaum wanted to invite Alice to the GDR to prepare for their panel for the WEHC in Bern in 1986, Nussbaum had to ask for approval from Czechoslovakia (Teich & Teichova, 2005, 217–218).

[6] Girton College Archive, GCPP Teichova 2/6/16. Academic correspondence of Alice Teichova and Mikuláš Teich: File 16. Kaplan to the Teichs, February 7, 1984; Mikuláš Teich to Kaplan March 5, 1984.

[7] Girton College Archive, GCPP Teichova 2/6/8. Academic correspondence of Alice Teichova and Mikuláš Teich: File 8. Myška to the Teichs, May 29, 1976.

[8] Interview with Jana Engelová, Česká Lípa, June 2021.

Until his final days, Mikuláš carried with difficulty not only his wrongful expulsion from the Party but also the stigma of being an exile. He used to say that a person is not only what he or she feels about herself or himself but also what other people think of him or her. He believed that in England he was considered Czechoslovakian. At the same time, he agonized deeply over the fact that their work was never given proper recognition in Czechoslovakia. He bore heavily the feelings they experienced when he and Alice came back to Czechoslovakia after the Velvet Revolution in response to an invitation from the Institute of History. Yes, I gave a lecture, he recollected bitterly, but I had a strong feeling after all that the people from Czechoslovak academia thought of us as people who emigrated and who were well off in Cambridge, so what we can to tell them![9]

Miroslav Hroch (1932): Not to Waste Any Opportunity!

When the ideologically uncompromising and pro-regime historian Čestmír Amort (1922–2013) was asked during his visit to Bulgaria in the 1970s whether he thought Miroslav Hroch was a Marxist, he replied: he is a Marxist, but of Yugoslavian style. Nevertheless, the methodology Hroch used in his analysis of nineteenth-century national movements is an accurate application of historical materialism. His world-known A-B-C theory of the national movement is based on a quantitative analysis of the economic and social assumptions of the national movement. Paradoxically, Hroch arrived the methodology thanks to his effort to avoid the Marxist-Leninist claims for the "progressiveness" of the national movement. He believed that hard figures were not vulnerable to ideological scrutiny and that social–historical empirical research free from parti-pris would depoliticize the topic (Hroch, 2018, 24). It turned out that although the study of national revolutionary traditions was in the core of Marxist-Leninist historiography, the study of the nation itself was not a conflict-free field of research. Hroch's forty-year career trajectory was firmly rooted in his time and influenced by the tightening and loosing of the screw of the communist regime in Czechoslovakia. Yet, and perhaps because of that, with one exception, he never missed an opportunity to collaborate across the Iron Curtain.

[9] Interview with Mikuláš Teich, Cambridge, May 7, 2018.

Miroslav Hroch was born in Prague. He studied at a classical gymnasium and, from 1951, literature and history at the Faculty of Arts at Charles University. Very soon history prevailed. Unlike literature that suffered from the 1948 purges, history seemed to him more factual, professional, and apolitical. After 1948, the teaching of non-communist teachers was limited, some even had to leave the faculty, but after 1953 some of them could return. As Hroch recollects, only during the last terms of his studies did the lectures on the history of the twentieth century look like political training (Hroch, 2023). Politicization penetrated also into the compulsory youth communist organization meetings full of flaming speeches and papers on the class fight or the Korean War, which were orchestrated by a few older students, fanatical communists who, later on, became famous for their participation in the Prague Spring movement and *Charta 77*. In the early 1950s, their only aim was to destroy the slightest remains of "bourgeois" elements in academia. With a bitter smile, Hroch still recalls one particular meeting where these students required all attendees to deliver a paper on Why I love Comrade Stalin and his fear that he would be given the floor.

An encounter with Josef Polišenský, his professor of history, was crucial for Hroch's later academic career. Polišenský voluntarily exchanged his Social Democratic Party membership for the Czechoslovak Communist Party in 1948 and became a crucial person in organizing the history curriculum at the Faculty of Arts; at the same time, he managed to secure his students in a bubble free from outside fanaticism and ideological clichés. Under his supervision, Hroch wrote his first academic papers on the social composition of supporters of the journal *Matice Česká* in the nineteenth century and also a master's thesis on the economic context of Wallenstein's military and political power during the Danish War. The later became in core of his PhD thesis, slightly transformed into an analysis of the relations between trade and politics (Hroch, 1963). As a PhD student, Hroch became an assistant librarian at the library of the Department of History. With the support of department heads who, despite their ideological bias, understood the need to maintain contact with "Western" science, he ordered foreign journals and books for the library and organized interlibrary loans with Scandinavian and some West European libraries.

Thanks to Polišenský's international prestige, Hroch met and had discussions with foreign experts from England, Scandinavia, and the Netherlands who visited Polišenský privately in Prague. He could also start traveling abroad—albeit only after Party approval and at first only to socialist countries: Budapest (1956) and Wroclaw (1957). In 1958, Polišenský faced attacks on positivism and revisionism in his work. Yet, as a chairman of the newly formed commission of historians of Czechoslovakia and the GDR, he initiated the exchange of eight assistants (including Hroch) with assistants from Humboldt University. In the same year, Hroch, while still a PhD candidate, attended for the first time the international scientific conference of Hansischer Geschichtsverein [Hansa Historical Society], one of the few and later the only scientific association to have members from both German states. Hroch presented the results of his thesis on Wallenstein's policy in the Baltic. In the following years, these conferences became literally a gateway to the world to him; there he got acquainted with what was happening in Western historiography and met a number of renowned but also young German (West and East), Dutch, Polish, Swedish, French, and Danish historians. Polish historians (Henryk Samsonowicz, Antoni Maczak, Maria Bogucka) were particularly close to him, and he later collaborated with them on the Czechoslovak-Polish Historical Committee. Thanks to Polišenský's relations with Swedish historian Sven Göransson, in 1958 Hroch set out on his first study trip abroad, to Sweden. The ideological recommendation he needed at the time was from a street committee of the Party. Neighbors who knew Hroch's mother sat on the committee. They told her in private that they had praised him and his family (Hroch, 2023).

While Czechoslovak scientists were still banned from cooperation with West German historians (this ban continued throughout the socialist period, with the brief exception of the mid-1960s), historians from other Western countries began to come to Prague in the early 1960s. Among them was also the historian who would be key to the future recognition of Hroch's work in the West, Eric Hobsbawm. The head of the department asked Hroch to guide Hobsbawm during his visit. Hroch remembers that Hobsbawm wanted to visit the Semafor Theater, which was then a Prague cultural phenomenon and its founder, Jiří Suchý, is still a respected artist in the Czech Republic. But Hobsbawm, who did not tolerate courtesy detours, said in the middle of the performance that Suchý was just an average singer. That made Hroch upset. He tried to explain that people went there for the lyrics, but Hobsbawm, who was a serious jazz enthusiast, dismissed him. "A British snob," Hroch thought of him then (Hroch, 2023).

Although Hobsbawm was a Marxist and even a member of the Communist Party of Great Britain, (Evans, 2019) he was not given the opportunity to lecture to students, only to teachers. This also applied to other British and French historians, including Ferdinand Braudel, who opened up collaborations with historians from the Annales school. Their protagonist in Prague was the head of the department, Václav Husa (1906–1965), who received scholarships for his assistants for study abroad in France and connected members of the department with an international project studying prices and wages. A number of Czechoslovak historians gradually became involved in the research (Dolezalova & Holec, 2016a).

The door to cooperation with West Germany was not opened to Hroch until 1963 with a visit by students from the University of Marburg. They were led by Peter Scheibert, and Hroch became their guide again. Of course, he also used the time to discuss his work. He had recently defended his PhD thesis and was considering a new field of research: a comparative analysis of social composition of national movements. After a heated discussion in the U kotvy pub, Scheibert shouted: "Herr Hroch, Sie müssen zu uns kommen!" Scheibert soon invited Hroch to Marburg and even secured funding for his stay on a Humboldt Foundation scholarship. But that was a problem. Czechoslovaks could not accept Western, let alone German, scholarships. Scheibert, therefore, applied directly for a scholarship. Hroch just asked his faculty for permission to go to Marburg. Faculty and the Party organization at the Faculty approved the task; so did the Ministry of Education. It got complicated when the Humboldt Foundation asked Hroch to send his diploma certificate. To provide this kind of information to the Western institutions was strictly prohibited in Czechoslovakia. Hroch sent the documents without permission (Hroch, 2023).

The three periods out of academic terms he spent in Marburg between 1964 and 1966 were decisive for his future research and career. Not only did he met Hungarian medievalist-in-exile János Bak and Slovak economic historian and future émigré in Darmstadt, Ákoš Paulinyi. Hroch also developed close friendly relations with Gottfried Schramm, the son of that time famous West German historian Percy Schramm. He still recalls discussions with him as an intellectual pleasure and inspiration. For the first time, he experienced in Marburg the atmosphere of focused and open discussion; just books, university-related issues, academic work, and some contemporary politics. Hroch was fascinated by the interdisciplinary approach governing discussion: students and professors of political sciences, sociology, and philosophy discussed together. Because the majority of the professors considered Marxism "salonfähig," it was in Marburg that Hroch started to take Marxism seriously as a methodological tool and claimed himself as a Marxist.

There was another permanent asset of Hroch's illegal Humboldt scholarship in Marburg. Originally, Hroch intended to address in his comparison the nations whose languages he had partially learned during his research on Baltic trade. While Hroch failed in his domestic application to travel to the Soviet Baltic republics to study in archives there, in Marburg he found Baltic scientists who helped him translate the books and contact their colleagues at home who focused national issues: German-Lithuanian student Helene Grauduschus, Soviet-Estonian Ea Jansen, and Finnish historian Aira Kemiläinen. Thus, in Marburg can be found the explanation of why Hroch addressed Finns, Norwegians, and Baltic nations in his comparison and not Welsh, Irish, or Flemish ones. He eventually included the Welsh in his analysis thanks to his student in Prague, Eva Vosáhlová, who expressed her interest in writing a master's thesis on the fight for the language character of Welsh elementary schooling. Hroch organized for her a scholarship at the University of Swansea in 1967. She learned kymrik, wrote and defended her thesis in Prague in 1968, and then decided to emigrate to Wales because of the Soviet occupation (Hroch, 2023).

Last but not least, in Marburg Hroch could read new literature and find inspiration: while sociology and comparative analysis, labeled "comparatism," was to the ideological powers a "bourgeois pseudo-science," Hroch explored the comparative method in the work of Dutch social psychologists H. C. J. Duijker and N. H. Frijda (1960). What, however, inspired and encouraged Hroch the most was the French series *L'histoire des civilisation* and Eric Hobsbawm's *The Age of Revolution* (Hroch, 2018, 23). He was also attracted by the ongoing discussion on the crisis of seventeenth century. He was well informed about the discussion from the Polish colleagues. In Marburg, however, he had an opportunity to read the entire spectrum of contributions to the discussion published in books or journals.

This influenced Hroch's return to his research on the trade between Western and Eastern Europe during the Thirty Years Wars that he had elaborated already in his master's and doctoral theses. In the early 1970s, he enlarged his explorations through archival research in Poland (Hroch,

2023) and focused on the state of the historiography and the transfer of ideas between Western and Eastern historians on Baltic trade; his analysis became the first part of the book on the crises of feudal society (Hroch & Petráň, 1976). On the one side, Hroch fully acknowledged the foundational role of Western historians in the discussion when he analyzed various approaches to the phenomena from R.H. Hilton's article published in *Annales* in 1950 (Hilton, 1950) until what was then the current state of the historiography: papers published in the first years after the 1970's congress in Moscow. On the other side, his point of departure was Marx's concept of crisis. Thus, he criticized Western historiography because it separated the problem of the seventeenth-century crisis from both the revolutionary change in societal formations and the transition from feudalism to capitalism. According to Hroch, historiography did not see the phenomena in its complexity because historians on both sides of the Iron Curtain did not take into consideration the findings of historians from the other side. As a consequence, the discussion was only limited and did not reach a generally accepted definition of crisis; moreover, historians, including Soviets, did not struggle to establish a generally valid definition, and their considerations had only weak theoretical foundations.

In his search for a valid definition, Hroch applied the concept of social and economic contradictions introduced by Marx in *Capital*: the crisis is a point of exacerbation of these contradictions and a departure from the existing balance in the prevailing social order. In the next step, Hroch verified in which spheres of social life (international politics, domestic political relations, technological and demographic development, trade, and agriculture, industry and crafts) this definition was fulfilled (Hroch & Petráň, 1976, 308–315). He found escalating conflicts in many areas of life in many parts of Europe at that time and pointed out that the crisis could not be studied without a focus on the connections between various factors: international relations could not be understood without economic development in individual countries, nor art without the influence of economic and social factors, nor domestic politics without international trade. He arrived at the definition of crisis as a sign of two aspects of historical development in the seventeenth century: (1) the accumulation of partial crises in various sections of economics, politics, and culture and (2) the impact of crisis phenomena in various spheres of social life (ibid., 295–296).

After its Czech edition, the book was translated into German, and mainly thanks to the help of Gottfried Schramm published in Hamburg (Hroch & Petráň, 1981), yet it did not stimulate any discussion. Hroch himself left the topic already after its Czech edition of 1976 when he realized that Immanuel Wallerstein had published the first volume of his *The Modern World System* (1974). Wallerstein completed what I only intended to do, Hroch recollects (Hroch, 2023).

Fortunately, at the same time, another opportunity arose to allow Hroch to step out of the bounds of the heavily guarded socialist science. When he published the outcomes of his research on the social preconditions of national movement in Czech, it did not attract the interest of Czechoslovak academia. However, he did grab the unforeseen opportunity to publish his book in German, after shortening and translating, with Charles University publisher Karolinum (Hroch, 1968). In the mid-1970s, the book reached German and British audiences thanks to the favorable reviews published by *Historical Review, American Historical Review, Vierteljahrschrift for Wirtschafts- und Sozialgeschichte, Annales,* and *Slavic Review.* The A-B-C phases proved beguiling to readers (Hroch, 2018, 32). As a consequence, the British publisher New Left Books proposed to publish the book in English translation. As at the time approval from the Party authorities was required for publishing in the West, Hroch was quite taken aback to learn that the publisher's profile included the name Trotsky and some Yugoslav revisionists. He wrote Hobsbawm to share his worries. Hobsbawm contacted Cambridge University Press, which agreed to publish an even broader version of the German version of the book. Very soon, however, political-bureaucratic obstacles arose.

Czechoslovak scholars were not allowed to send manuscripts abroad privately for publication. The only legal path involved the state agency Dilia, which held a monopoly on representing intellectual property rights outside the country. Dilia required approval from the author's faculty and an endorsement guaranteeing that the text was politically and ideologically suitable for publication in the West. Since the publisher signed the contract after receiving approval from the faculty, Hroch stopped trying to obtain the endorsement. When the Czech manuscript was ready in 1980, Dilia sent it to the publisher without asking for the endorsement. In a year and a half, after the book was translated, CUP asked Hroch to shorten the manuscript. After more than a year of shortening and rewriting, Dilia shipped the new manuscript to CUP in 1983. It never reached the publisher! Accidentally, when Hroch attended a conference at the School of Slavonic and East European Studies in London in 1984, he shared his bitter experience with Robert Pynsent. Pynsent telephoned Cambridge immediately, and Hroch traveled to Cambridge to get another copy of the translation. Then he had to repeat the same time-consuming procedure. The book was published, finally, in 1985 under the title *Social Preconditions of National Revival in Europe* (Hroch, 1985). After the Velvet Revolution, Hroch visited the officer in Dilia and asked her whether she knew he had no endorsement. She replied with laugh, of course I knew (Hroch, 2018, 34–36).

Despite his manifold relations with his foreign colleagues, Hroch only rarely took part in large international congresses. Partially because he always preferred small workshops and intensive contacts in intimate atmosphere. Partially because he never was a preferred representative of Czechoslovak historiography. He was selected for a large congress only once: for overcrowded Moscow, where he contributed a paper to discussion on the crisis of the seventeenth century. Ten years earlier, he had been asked to compile an overview of Swedish historiography for the Czechoslovak delegation to the congress in Stockholm in 1960, using information from his study trip to Sweden in 1958. He visited only one WEHC and it was on his own resources; in Munich in 1965 when he stayed in Marburg and was financially supported by the Humboldt scholarship. From 1970s, even though his activities were partially shielded by his Party membership, he was not the most intensively traveling employee of his faculty.

From the 1960s, Hroch's travelings abroad were under close surveillance by the secret police. While in the 1960s he was asked to report on the West Germans' opinion of Czechoslovakia, requirements to report on meeting Czechoslovak historians in exile and their attitudes toward the communist regime prevailed in the 1980s. As Hroch recalls, this was why he carefully avoided meetings with émigrés until 1988 when he met the Teichs in Bielefeld. The vigilance of the regime, however, was not stable. When he was driving home from Sweden in 1958, Emil Schieche of the University of Uppsala asked him to pick up books at the Collegium Carolinum in Munich and bring them to Prague. The books filled two pieces of luggage. When the Czechoslovak custom officer at the border asked Hroch about the content of these suitcases, he answered, books. And the officer said the books are custom free, you can continue. When he catalogued these books in the history department library, Hroch surprisingly realized that books were on the expulsion of the Sudeten Germans after World War II, literature that was prohibited in Czechoslovakia at that time. When he was returning from Denmark in the 1980s, unlike the situation in 1958, the custom officer at the borders went through all of his papers from the conference searching for some suspicious material of sabotage (Hroch, 2023).

The success of *Social Preconditions* enabled Hroch to re-establish his contacts with western colleagues in the mid-1980s. He was invited to Hobsbawm's Wiles lecture in Belfast in 1985, where he encountered Perry Anderson and John Breuilly. He participated in two projects: in Bielefeld on the European process of *Verbürgerlichung* led by Jürgen Kocka, and in the project named Comparative Studies on Governments and Non-Dominant Ethnic Groups in Europe initiated by Gerald Stourzh, financed by European Scientific Foundation, and led by Adrea Keppler. Gale Stokes invited him for a conference in Washington to chair a discussion on Roman Szporluk's book on the concept of nation in the writings of Friedrich List and the young Karl Marx. When, however, Stokes invited him to a guest professorship at Rice University in 1988, Hroch refused the invitation because of his insufficient command of English. Yet, further invitations followed very soon as the Iron Curtain fell down in one year and the topic of nationalism became fashionable.

The Industrial Revolution Debate of Socialist Style

The perception of and research on the Industrial Revolution in the Central European environment developed in more ways despite the official narrative based on Marxism-Leninism. Discussions between representatives of individual socialist historiographies were full of tensions, professional conflicts, and personal animosities. It was a topic that they found difficult to squeeze into their ideological frameworks and hard to make pursuant to the classical lectures of Marxism-Leninism. Since the second half of the 1950s, the issue of eighteenth- and nineteenth-century economic development and the inherently connected Industrial Revolution became one of the passionately discussed topics in Czechoslovak economic history for several reasons. First, it aligned well with Marxist preferences for economic development in the historical process. Second, it presented a topic and a period that were less plagued with ideology than others. Various international conferences and seminars took place; in them Jaroslav Purš, a politically reliable and undoubtedly capable Czech historian, took over the role of arbiter. His ambition was also to step into international polemics and present his meritorious standpoints. He often commented on foreign literature via the *Review*, pondering on it creatively and perusing it even when re-formulating his own theories.

The first Czechoslovak polemics were caused by attempts of some Slovak historians, especially Anton Špiesz (1930–1993), to shift the beginning of the Industrial Revolution in the Slovak environment to the second half of the 1780s because of the extensive development of manufactures on the territory of contemporary Slovakia in the Theresian and Josephian periods. In his polemics with Purš, Špiesz relied too heavily on the British model while Purš accentuated the Central European specifics as well as the connection to nationality and social history issues. Purš's concept, with its broader international foundation and theoretical persuasiveness (steam engines as the attributes of the Industrial Revolution), dominated Czechoslovak discourse and evoked new discussions, this time with Hungarian historians. Initially, Slovak historians dominated the debate due to the Purš's language barrier, and Purš had to filter the findings of the polemics via Slovak colleagues or via translations of the Hungarian papers to other, more accessible languages, even though he officially pretended to have worked directly with the Hungarian sources.

After the Prague Spring ended, Špiesz gradually retreated from the discussions for his strong Catholicism. His presence at the Institute of History of the Slovak Academy was merely tolerated while his opportunities to enter international discourse were limited and purposefully clipped. It was only in 1971 that he returned to the topic by comparing the Industrial Revolution in England with the territory of the Bohemian lands and the Upper Hungary, the present Slovakia. Although he came back to rely on the British model based on the most recent British literature in the field (H. Heaton, P. Deane, Ch. Hill, W. O. Henderson), previously criticized by Purš, when it came to the development in the Bohemian lands, he was ready to accept Purš's conclusions. When regarding Slovakia, however, he left the proverbial door open to interpretation (Špiesz, 1971, 68). The methodologically incorrect tradition of removing of the Bohemian lands from the Austrian economic context and Slovakia from the Hungarian contexts prevailed in Central Europe as the result of both the dissolution of the united economic space after 1918 and the national approaches of individual national historiographies in the area.

Contrary to Špiesz, Arnošt Klíma was able to cooperate more tightly with the international research of the time, being a member of the Executive Board of the IEHA. Despite his international reputation, however, Klíma became a *persona non grata* in Czechoslovakia after the Prague Spring and had to face the displeasure of the communist regime. With the assistance of Alice and Mikuláš Teich, he had published his paper in the renowned *Past & Present*, making Klíma the only East European historian participating in a crucial international discussion on the genesis of capitalism, evoked by a controversial study of Californian professor Robert Brenner from 1976 (Klíma, 1979). The discussion proceeded on the pages of the Marxist-prone journal *Past & Present* until the beginning of the 1980s. It also included Malthusian concepts and opinions of the significant French historian Emmanuel Le Roy Ladurie from the Annales School and came to an end in 1982 via Brenner´s study on the agrarian roots of European capitalism. All of the commentaries were collected in a single published document, which was later republished a few times (Aston & Philpin, 1987).

In the meantime, due to his social, political, and scientific authority, Jaroslav Purš became a leader of the home discussion with his own detailed concept of the Industrial Revolution, which he was persistently inserting into officially accepted texts. He pushed his theses in an uncompromising way, though not without considerable scientific creativity. He was slowly building his argumentative foundation via numerous analytical and synthetic works and summarized them in a monumental and awe-inspiring work *Průmyslová revoluce Vývoj pojmu a koncepce* [*The Industrial Revolution. Development of the Term and Concept*] (1973). He became a professor at Charles University and in 1976 also the director of the Institute of Czechoslovak and World History of the Academy in Prague and was considered an overall undisputed opinion-forming authority in Czechoslovakia. He reinforced his position through his linguistic skill set, opportunities to travel, and forming his own historic school, handpicked from his postgraduate students, who were given partial research tasks to work on regarding problems and parallel processes of the Industrial Revolution. Other Czech and Slovak historians writing about the same topic did not find a way to free themselves from his concepts, which became part of the all-state syntheses. Purš finally evolved to embrace a model of asynchronous peaks of the Industrial Revolution related with the first cyclical capitalist crises, as well as the phase shift of individual social processes, as the regions of future Czechoslovakia with their territorial differences seemed to him a satisfactory model laboratory.

In Purš's synthesis of 1973 the name index is done in a very sloppy way, which makes one doubt whether it was not deliberate. Some historians or their most significant works were omitted: a neglected Czech economic historian and economist living in Slovak "asylum" Jozef Faltus (1929–1999) was mentioned only fleetingly. The originally Slovak exile historian of science and engineering Ákos Paulinyi and the pro-regime Slovak economy historian Pavel Hapák (1931) were completely missing. Such a selective approach became the target of strong criticism by Jürgen Kuczynski of the GDR, who saw Purš as ignorant of a number of works from the socialist environment. In fact, Kuczynski became Purš's only opponent in the eastern part of Europe. Similar professional, linguistic, and political skills allowed Kuczynski to play a crucial role in East German historiography. A ruthless conceptual fight broke between the two historians standing at the helm of key academic institutions and having remarkable reputations, leading to a wide range of animosities, even in their personal lives. Conflicts can be traced down in international speeches as well as in published works in various languages.

The beginning of different approaches and mutual controversies goes back to 1956 when, after visiting Prague in 1954, Kuczynski got acquainted with the Purš's work. Their polemics focused on the process of the Industrial Revolution in the conditions of the so called "Prussian Way"of the East European agricultural development, on the topics of chemical technological processes and applications of scientific knowledge in the Industrial Revolution. Nevertheless, they also debated the narrow focus of the Industrial Revolution on the cotton-making industry or silk production. They also discussed the alleged peak of the revolution in the German environment only by the end of the nineteenth century, the role of the food processing industry in the lands east of the Elbe river, their differences in how they applied Marxist and Leninist interpretations, as well as the simple setting of individual processes into timelines. Purš time-phased revolutionary processes in the economy deep into the twentieth century and conditioned them by individual dominant factors. This caused interdisciplinary crossovers with the works of other professionals dealing with the history of the twentieth century.

Despite the extent of the Purš's synthesis, many concepts remained untouched or only slightly analyzed. He did not participate in the discussion on the genesis of capitalism and the concept of proto-industrialization. The idea of late industrialization and the problem of its Central and East European specifics was not in his scope of interest, even though this was obviously a productive direction for the region and for Czechoslovakia which was crippled by big geographic and economic disproportions. That became the professional turf of the renowned economic historian Milan Myška from Ostrava.

Myška, who had been shunned by university circles since 1970 for his refusal to accept the Warsaw Pact troops' invasion of Czechoslovakia as a liberation, struggled throughout the normalization period living without a permanent job. Even though he could only publish at home under another name (mainly invented such as Tomáš Pavlica, Pavla Bílková, Jitka Noušová, Vlastimila Čechová), he never resigned from conceptual historical research. Thanks to both the international contacts he had built as a scientist and academic official before 1968, and his friendship with Polish historians and their access to international literature, he was well versed in the contemporary research field and on the receiving end of many German and Polish-mediated sources. Paradoxically, Klíma and Myška in their internal exile had a wider range of opportunities than the official historians, who were limited by the decisions of the Party and by their social and political roles. The latter had to come to terms with a bigger isolation from research than the so-called sideline historians; the voice of whom was heard louder abroad and who could, generally speaking, benefit much more from international cooperation. This is shown in the access to various literature and, in case of Klíma, also personal contacts.

As a result of receiving such mediated information, Myška started working on a new scientific paradigm spreading from western Germany, the so-called proto-industrialization concept. It offered a much better transition from the manufacture period into the phase of the Industrial Revolution, including all industrial sectors. Like Klíma, Myška benefited from the help of Alice and Mikuláš Teich and thus came to be seen as an important authority on the topic (Myška, 1979). When in 1982 the economic history congress took place in Budapest, Myška received a personal invitation. The congress president, György Ránki, issued the invitation, probably after the intervention of Klíma who at that time was a guest professor at the university in Bielefeld. Since Hungary was perceived as a neighboring and comrade socialist country, Myška was able to travel there without any hindrances. He took active part in the congress discussions within the proto-industrialization section. Kuczynski openly regarded him and his contribution as valid, while the leader of the official Czechoslovak delegation, Purš, did not even mention him. Purš considered the concept of proto-industrialization as anti-Marxist and never accepted it (Purš, 1983).

In 1987, Myška dared to reprimand the then-omnipotent Purš (though only on the pages of regional periodicals) for not considering more specifics of the Industrial Revolution, as they were taken into account successfully through the lens of the lagging development of Central and Eastern Europe in the works of Alexander Gerschenkron and Hungarian historians Iván T. Berend and György Ránki. Moreover, Myška saw the setting of Purš's timeline for the Industrial Revolution from the verge of the seventeenth to the eighteen century until the 1870s as far too optimistic (Myška, 1987). Opposed to Purš, who had relied rather on British literature, Myška and his work could be characterized by his connection to Polish colleagues, mainly W. Kula and W. Długoborski, who transmitted the results of the French Annales school. They also could offer methodological stimuli because they dealt with historical conditions very similar to those in the Bohemian lands, especially Silesia.

In a different realm and as a true lone wolf, there was another historian (of engineering) Ákoš Paulinyi who contributed to the discussions on the character of the Industrial Revolution, even though not in the framework of Czechoslovak historiography and not really with it. His papers shifted attention to the role of technology in the process of the Industrial Revolution. In 1969, Paulinyi left Czechoslovakia and completely assimilated to the German environment not only bodily but also mentally. He kept his long-term friendly relations with the Teichs and together they all refused contacts with the political emigration and its anti-regime activities (although they had different reasoning); Paulinyi also completely severed all his connections to Czechoslovakia and colleagues at home.

To summarize, the official direction of Czechoslovak historiography in research on the Industrial Revolution was represented by Purš and his school. The unofficial, yet by no means weaker, segment was represented by excellent historians who were either sidelined by the regime or persecuted in various ways and who, based on their subjective situation, set the boundaries of their own internal exile (Klíma, Myška). The second ones read the works of the first ones, while the first ones made sure to stay away from the second ones. Somewhere in between both camps were domestic historians who did not represent the *crème de la crème* of the regime yet were enabled to do research and express their opinions, such as Špiesz, Horská, and Ladislav Tajták (1921–2020). However, their possibilities to participate in international discussions were limited, especially by the clipped or completely prohibited travel options. The fourth segment was represented by exile (Paulinyi, Teich). Their approach to the historiography at home was far from united or one-sided. While Paulinyi ostentatiously ignored Czechoslovak historiography, the Teichs attempted to help and cooperate. Potentially bigger possibilities of international discourse were diminished by difficult access to Czechoslovak sources, which applied as well to the historians at home. It must have been complicated to do research and formulate conclusions when both sides kept on passing each other by without encounter.

Unfortunately, when the above-mentioned historians left and the school of Purš dispersed after 1989, theoretical analysis of such concepts as the Industrial Revolution, proto-industrialization, and partially also economic modernization in Czech and Slovak historiography remains on the periphery of interest.

Conclusion

The trajectory of the historiography of economic history in Czechoslovakia was affected by geopolitical changes, the undemocratic totalitarian regime and ideological pressures as well as by the waves of emigration. International cooperation was an ambiguous aspect of the trajectory. As the analysis of the institutional framework for international cooperation reveals, international cooperation and travels abroad were not forbidden, but were highly selective and political issue connected with ideological and bureaucratic obstacles. These obstacles were created and carefully observed by the same institutions that managed scientific research and provided resources; obstacles were aimed toward individuals not toward the concept of cooperation. The selection of participants for international events and travels abroad was subordinated to Party decisions.

At the same time, economic historians were regularly updated with new developments in world historiography through the main home journals and leading journals in the field that were available in the main libraries in the country. As the case study on the Industrial Revolution demonstrates, Czech and Slovak historians were inspired by various streams of East and West historiography, and their voices were not unanimous. The experience of Teichova and Hroch reveals that the Iron Curtain was not an impenetrable borderline but mainly a state of mind and a way of thinking. For Teichova and Hroch, as they strove for scientific truth, international cooperation was a way to maintain academic quality and scientific openness regardless of borders.

References

Archiv bezpečnostních složek, Collection Archivní protokol operativních svazků centrálních útvarů MV a KS MV [Archive of the Security Services. Collection Archival Protocols of the Operational Units of the Central Departments of the Ministry of the Interior and Regional Directorates in Prague].

Girton College Archive, Collection GCPP Teichova.

Aston, T. H., & Philpin, C. H. E. (Eds.). (1987). *The Brenner Debate: Agrarian Class Structure and Economic Development in Pre-industrial Europe*. Past & Present Publications. Cambridge University Press.

Aston, T. H., & Hilton, R. H. (1974). Historians in Czechoslovakia. *Past & Present, 63*(1), 3. https://doi.org/10.1093/past/63.1.3

Berg, M. (2015). East-West Dialogues: Economic Historians, the Cold War, and Détente. *Journal of Modern History, 87*(1), 36–71.

Doležal, K. (2022). Československý poválečný nacionalismus v mediální recepci snímku Jan Roháč z Dubé. *Historický Časopis, 70*(1), 87–114.

Doležalová, A. (2016). An Economic Historian of Central Europe. *Austrian History Yearbook, 47*, 191–194.

Doležalová, A. (2018). *A History of Czech Economic Thought*. Routledge.

Doležalová, A. (2021). *Mikuláš Teich (1918–2018): Moje století. Intelektuální biografie v dialogu*. Galén.

Doležalová, A., & Holec, R. (2016a). Continuity and Discontinuity in the Czech and Slovak Historiographies. In P. Hudson & F. Boldizzoni (Eds.), *Routledge Handbook of Global Economic History* (pp. 230–241). Routledge.

Doležalová, A., & Holec, R. (2016b) Vplyv exilových hospodárskych historikov na československú historiografiu. In *Česko-slovenská historická ročenka* (Vol. 2014–2015, pp. 143–152). VEDA.

Duijker, H. C. J., & Frijda, N. H. (1960). *National Character and National Stereotypes: A Trend Report Prepared for the International Union of Scientific Psychology*. North-Holland.

Erdmann, K. D. (2005). *Toward a Global Community of Historians*. Berghahn Books.

Evans, R. J. (2019). *Eric Hobsbawm: A Life in History*. Oxford University Press.

Hobsbawm, E. (1962). *The Age of Revolution: Europe, 1789–1848*. Weidenfeld & Nicolson.

Hroch, M. (1963). Der Dreissigjährige Krieg und die europäischen Handelsbeziehungen. *Wissenschaftliche Zeitschrift der Ernst-Moritz-Arndt-Universitäsdt Greifswald, XII*(5/6), 533–543.
Hroch, M. (1968). *Die Vorkämpfer der nationalen Bewegung bei den kleinen Völkern Europas: eine vergleichende Analyse zur gesellschaftlichen Schichtung der patriotischen Gruppen*. Karlova univerzita.
Hroch, M. (1976). *Handel und Politik im Ostseeraum während des Dreissigjährigen Krieges: zur Rolle des Kaufmannskapitals in der aufkommenden allgemeinen Krise der Feudalgesellschaft in Europa*. Univerzita Karlova.
Hroch, M. (1980). *Das westeuropäische Kaufmannskapital als Faktor der wirtschaftlichen Differenzierung von Ost und West in der früheren Neuzeit*. Institute of Czechoslovak and World History of the Czechoslovak Academy of Sciences.
Hroch, M. (1985). *Social Preconditions of National Revival in Europe*. Cambridge University Press.
Hroch, M. (2018). *Studying Nationalism Under Changing Conditions and Regimes. An Intellectual Biography. Essays 3*. Peristyle.
Hroch, M. (2023). *Jak jsem to viděl* (Forthcoming).
Hroch, M., & Petráň, J. (1976). *17. století – Krize feudální společnosti*. Svoboda.
Hroch, M., & Petráň, J. (1981). *Das 17. Jahrhundert – Krise der Feudalgesellschaft*. Hoffmann und Campe.
Jiroušek, B. (2006). *Česká a československá účast na mezinárodních kongresech historických věd*. Jihočeská univerzita.
Jiroušek, B. (2008). *Proměny diskursu české marxistické historiografie*. Jihočeská univerzita.
Kaplan, K. (1986). *Die politischen Prozesse in der Tschechoslowakei 1948–1954*. Oldenbourg Verlag.
Kaser, M., & Radice, E. A. (Eds.) (1985/1986) *The Economic History of Eastern Europe 1919–1975*. Oxford University Press.
Kladiva, J. (1961). XI. mezinárodní kongres historiků ve Stockholmu. *Československý časopis historický, 9*(1), 151–162.
Klíma, A. (1957). Industrial Development in Bohemia, 1648–1781. *Past & Present, 11*(1), 87–99.
Klíma, A. (1974). The Role of Rural Domestic Industry in Bohemia in the Eighteenth Century. *Economic History Review, 27*(1), 48–56.
Klíma, A. (1979). Agrarian Class Structure and Economic Development in Pre-industrial Bohemia. *Past & Present, 85*(1), 49–67.
Klíma, A. (1986a). Agrarian Class Structure and Economic Development in Pre-Industrial Bohemia. In T. H. Aston & C. H. E. Philpin (Eds.), *The Brenner Debate: Agrarian Class Structure and Economic Development in Pre-Industrial Europe* (pp. 192–212). Cambridge University Press.

Klíma, A. (1986b). Die Arbeiterunruhen in Böhmen 1844. In H. Reinalter (Ed.), *Demokratische und soziale Protestbewegungen in Mitteleuropa 1815–1848/49* (pp. 230–264). Suhrkamp.

Macek, J. (1953). (1953) Mezinárodní kongres historiků v Budapešti. *Československý Časopis Historický, 1*(3), 410–416.

Myška, M. (1979). Pre-industrial Iron-making in the Czech Lands: The Labour Force and Production Relations Circa 1350—Circa 1840. *Past & Present, 82*(1), 44–72.

Myška, M. (1987). Opožděná industrializace. Protoindustriální výrobní vztahy a formy za průmyslové revoluce na příkladu lnářského průmyslu ve Slezsku. In *Z dějin textilu. Studie a materiály 11* (pp. 95–159).

Olšáková, D. (2014). *Věda jde k lidu! Československá společnost pro šíření politických a vědeckých znalostí a popularizace věd v Československu ve 20. století*. Academia.

Polišenský, J. (2001). *Historik v měnícím se světě*. Karolinum.

Prečan, V. (1994). *V kradeném čase*. Doplněk.

Purš, J. (1963). Druhá mezinárodní konference o hospodářských dějinách. *Československý Časopis Historický, 11*(2), 283.

Purš, J. (1983). VIII. mezinárodní kongres hospodářských dějin v Budapešti 1982. *Československý časopis historický, 31*(4), 631–640.

Sommer, V. (2011). *Angažované dějepisectví*. Nakladatelství Lidové noviny.

Špiesz, A. (1971). Priemyslová revolúcia v Anglicku a na území dnešného Československa (Pokus o porovnanie). *Historický Časopis, 19*(1), 49–69.

Teich, M, & Porter, R. (Eds.) (1986). *Revolution in History*. Cambridge University Press.

Teich, M., & Porter, R. (Eds). (1992). *The Scientific Revolution in National Context*. Cambridge University Press.

Teichová, A. (1968). *Dějiny středověku*. SPN.

Teichová, A. (1974). *An Economic Background to Munich: International Business and Czechoslovakia, 1918–1938*. Cambridge University Press.

Teichová, A., & Hánl J. (1960). *Dějepis pro 2. ročník odborných škol*. SPN.

Teichová, A., & Teich, M. (2005). *Zwischen der kleinen und der großen Welt: Ein gemeinsames Leben im 20. Jahrhundert* (G. Dressel & M. Reischitz, Eds.).

Wallerstein, I. M. (1974). *The Modern World System. 1. Capitalist Agriculture and the Origins of the European World-Economy in the Sixteenth Century*. Academic Press.

CHAPTER 5

Hungary: Surfing Among Ice Floes

György Kövér

During World War II two university departments operated in Hungary with the task to research and teach economic history: the Department of History of Hungarian Civilization at the Pázmány Péter University chaired by Sándor Domanovszky (1877–1955) and the Department of Economic History and Auxiliary Historical Sciences at the József Nádor University of Technology and Economic Sciences, led by Ferenc Kováts (1873–1956). Both professors were internationally known specialists of medieval economic history; moreover, Domanovszky was a school-founder personality. He drove a lot of students in the direction of

This paper has been supported by the research group "A professzionalizáció története Magyarországon a 19–20. században európai kontextusban" [*"The History of Professionalization in Hungary in the 19th and 20th Centuries in a European Context"*] (NKFI FK 132451).

G. Kövér (✉)
Eötvös Loránd University, Budapest, Hungary
e-mail: kover.gyorgy@btk.elte.hu

A. Dolezalova and C. Albrecht (eds.), *Behind the Iron Curtain*, Palgrave Studies in Economic History,
https://doi.org/10.1007/978-3-031-31578-7_5

the history of commerce (especially of foreign trade), or toward agricultural history (especially manorial economy). The best students of Domanovszky became associate professors (***Privatdozent***) at the universities: Miklós Komjáthy (1909–1993) and Imre Wellmann (1909–1994) at the Pázmány University and Jenő Berlász (1911–2015) at the József Nádor University. The best students were selected for continuation of a living tradition of economic history in Hungary after World War II (Kövér, 2016). However, due to the communist takeover this original plan failed. Domanovszky had been pensioned off and the Department of History of Civilization ceased to exist, together with the institution of ***privatdozentur***, and the designated successors could continue their career only as archivists or librarians. For the key persons of the younger generation of the Hungarian economic and social history in Transylvania, Zsigmond Jakó (1916–2008) and István Imreh (1919–2003), economic history study was a restricted but existing possibility to survive in their profession. Instead of the Faculty of Economics, in Budapest a new University of Economics was founded, which in 1953 received the name "Karl Marx." The chair of the new Department of Economic History at the university became Pál Zsigmond Pach (1919–2001), who originally had written his diploma work for Domanovszky (Kövér, 2020). This department became the cradle of the new Marxist economic history. The other flagship of Hungarian economic history writing was the Institute of History at the Hungarian Academy of Sciences (hereafter Institute of History at the Academy), where Pach was in the position of deputy director (until 1956).

For economic history writing new perspectives opened under the new institutional circumstances. Marxism proclaimed that the economic base, productive forces, and relations of production were determining historical development. Everyone following an academic profession had to adapt to the ideological clichés of Soviet-type Marxism: in the so-called system of academic qualification, everybody had to take an examination on Marxism; moreover, there was a strict ideological reviewing mechanism concerning the Marxist spirit of the interpretation of sources.

At the same time, in every period the current economic policy of the Hungarian Socialist Workers' Party (until 1948 named Hungarian Communist Party, between 1948 and 1956 Hungarian Working People's Party, hereafter the Party) determined the choice of preferred topics. During the Stalinist industrialization in the 1950s, industrial history enjoyed preference. Meanwhile, in the spirit of class struggle, heavy efforts were made at the Marxist reevaluation of the history of the peasantry. After the 1956 revolution, during the second wave of the collectivization of peasant plots, an *Agricultural History Review* was founded to investigate the historical relics of the just-liquidated peasantry. Since 1906 economic history had not had its own periodical. Between 1958 and 1975 the Department of Economic History of Karl Marx University edited a book series with the title *Treaties of Economic History*, which consisted of seven volumes on agricultural, industrial, commercial, and urban history issues. Due to the inspiration of the so-called new economic mechanism in 1968, which recognized and introduced market factors into the regulation of economic policy, new economic history research turned not only toward the reinvented entrepreneur, but also toward the historical analysis of company structure. Research into the international embeddedness of the Hungarian economy in the framework of the neo-Marxist core-periphery relations theory gained ground as well. After 1956 some old specialists who had not turned their backs on economic history or had not changed topic and period received the possibility to return, not to higher education, but rather to the Institute of History of the Academy, archives, and museums. After the crushing of the 1956 uprising every important historical debate, directly or disguised, treated the question of national identity, as did the sole, at least seemingly economic history discussion concerning the possibilities of the so-called autonomous capitalist development during the period of the Austro-Hungarian Monarchy (Kövér, 2009). By the 1980s, due to the intensification of international contacts and coming on the scene of a new generation of economic historians, the former rigidity of government politics regarding science became more elastic and colorful.

This study answers the question what were the options for international cooperation with Western historians from 1948 to 1990 in Hungary and what was the influence of international cooperation on the trajectory of Hungarian economic history. Special attention is dedicated to Hungary as an active organizer of international networks of economic historians and the only World Economic History Congress that took place in East European countries outside of the Soviet Union.

The Asymmetric Flow of Persons and Ideas

After World War II in Hungary the field of economic and social history was not considerably afflicted by emigration. For this reason, we focus first on the asymmetric migration of concepts. The 1950s are frequently referred to as a period of freeze in terms of the flow of ideas. To what extent this implied complete isolation is difficult to appraise.

Relying on a single parameter—library catalogs—I will attempt to show the extent to which Western economic history journals were available in Hungary's libraries. The library of the Institute of History at the University of Debrecen, the University Library of Eötvös Loránd University (ELTE), and the library of the Institute of History at the Academy each carried the journal *Annales* continuously. Except for a few years, it was also carried by National Széchényi Library, the library of the Academy, and the library of the University of Pécs. However, the journal *Vierteljahrschrift für Sozial- und Wirtschaftsgeschichte*, which was the oldest social and economic history journal and probably the one that had traditionally been the most widely read in Hungary, was carried continuously only by the University Library of Eötvös Loránd University and by the library of the University of Szeged. A complete series of the American *Journal of Economic History* was only available in the University Library of Eötvös Loránd University, and no complete series of the British *Economic History Review* was available anywhere in the country; even the University Library of Eötvös Loránd University is missing the complete years of 1950 and 1952. Without jumping to far-reaching conclusions from these facts, we may safely establish that the opportunities for keeping abreast of intellectual trends were better than we usually tend to assume.

The transfer of ideas was inhibited by a number of other factors instead. From 1950 on, the Institute of History at the Academy published *Értesítő*, a newsletter with the primary mission of publishing Soviet papers in Hungarian translation; however, it also printed regular press reviews. The very structure of the review is typical of the era. The first issue only featured a digest of the titles of papers published in the journals of the Soviet Union and the satellite people's democracies, but it was also announced that from the second issue on reviews of articles "from under imperialistic rule" would also follow; as explained by the publisher, "partly in order to ensure that we can report on the achievements of progressive history writing in those countries, and partly in order to be aware of the ambitions of bourgeois-imperialist history writing, our enemy in the battleground of our ideological struggle" (*A Történettudományi Intézet Értesítője*, 1950, 1).

But as far as economic history journals are concerned, only two French journals (*Annales*, *Revue d'histoire économique et sociale*) and one British journal (*Economic History Review*) made the selection. By the middle of the 1950s, the list was expanded to include the *Journal of Economic History* from the United States, but *Vierteljahrschrift* continued to be neglected (obviously, by choice). As far as the acquisition of new books is concerned, we have unfortunately no statistics on the deficiencies or discontinuities of that process.

In addition to personal contacts during the various conferences and congresses, another important vehicle of the opening toward the West was long-term grants involving traveling to and working in foreign countries. After 1956, Hungarian foreign policy intended to preserve its totalitarian nature while demonstrating a more open attitude in its foreign relations. Hungary made efforts to normalize its bilateral relations and concluded cultural agreements with Austria, Finland, France, and Italy. In addition to the foreign policy autonomy (neutrality) of most of these specific countries, an equally important factor was the fact that Hungary had a tradition of cultural relations with them. On the basis of the French–Hungarian cultural agreement signed in 1961, about one-third of all bilateral intergovernmental grants was financed by the French government (Somlai, 2009, 283–285). A dedicated agency, Kulturális Kapcsolatok Intézete [the Institute of Cultural Relations], was put in charge of supervising international mobility; eventually, this institute also acquired the right to decide about the allocation of grants across the various Hungarian institutes operating abroad. Of course, bilateral agreements ran the risk of one of the parties pushing an agenda of reciprocity, something the countries of the Eastern Bloc were wary about not only because of foreign exchange restrictions but also because of a fear of "soft subversion." The extent of the opening is reflected in the fact that "while in 1960 a total of 863 Hungarian researchers travelled abroad, and the Academy hosted 878 guests, just five years later the records show 1,476 foreign trips and 1,689 visitors from abroad" (Takács, 2014, 1503).

The 1963 reopening of the Collegium Hungaricum on Hollandstraße in Vienna may be considered the renewal of the tradition that had existed between the two world wars. The festive nature of the occasion did not, however, resolve the underlying diplomatic antagonism. The Austrian Foreign Ministry took offense because the Hungarians had engaged in preliminary discussions with the Ministry of Education and not with the foreign affairs administration and refused to delegate a dignitary to the event. The rift within the Austrian government coalition hinged on the conflict between B. Kreisky of the *Sozialdemokratische Partei Österreichs* and H. Drimmel the *Österreichisch Volkspartei* (Murber, 2005, 30; Ujváry, 2017, 245–259). Apparently, this internal fault line had more relevance for the Austrians than the process of international cultural rapprochement. An average grant to Austria lasted one month.

A new opportunity came along with the visit of Fernand Braudel to Budapest November 13–20, 1962 (Czoch, 2015, 121). In negotiation with the Academy, Braudel offered an annual total of 24 to 36 months of research grants for the use of Hungarian social sciences, every individual researcher being offered a three-month grant to work abroad. It was a typical reflection of the era that this academic initiative irritated not only the head of the Department of International Relations at the Academy, who was an affiliate of the intelligence services, but also the cultural attaché of the French Embassy in Hungary, who felt side-lined as he had no say in the selection of those allowed to travel (Czoch, 2015, 133–135). Even though no Hungarian historians traveled in the first round, it did not take long before Hungarian economic historians such as a polymath in the field of historiography (agricultural, technological, urban, cultural, and ecclesiastical history) László Makkai (1914–1989), an economic historian (of early modern period) Vera Zimányi (1930–2021), and one of the Hungarian pioneers of historical demography and quantitative economic history László Katus (1927–2015)—all representing the Institute of History at the Academy—could also travel in 1965 (Katus, 2008, 453). The reverse flow of cultural transfer must also be mentioned: almost one hundred books (sent by Braudel) arrived at the library of the Institute of History (Czoch, 2015, 125), something that certainly caught the eye of the intelligence services, among other agencies.

Braudel's visit was part of his wider East European strategy aimed at creating institutional contacts, but its consequences reached far beyond what we have described in the previous passage. The Hungarian reception of the Annales school was designed and built in a systematic manner that is not evident in other trends in history writing. The presentations and publications of French scholars visiting Budapest were just one of many channels dedicated to these efforts. In the spring of 1966 Le Goff delivered a widely popular lecture at the Institute of History under the title "*Does the Annales School even exist?*" which was also published (Le

Goff, 1968). Equally importantly, a series of debate sessions took place dedicated specifically to the promotion and interpretation of the Annales school's approach to history writing. It was probably not an accident that the rapporteurs of the debate during the February 26, 1968 session were László Katus, László Makkai, and Emil Niederhauser (*Századok*, 1969, 1118–1134).

The pinnacle of the Hungarian reception of the Annales school was the first French–Hungarian economic history conference, the French presentations of which were also published in Hungarian in *Világtörténet* [*World History*], a journal only available in camera-ready copies at the time (Világtörténet, 1968, 16. 3–73). The study of long-term economic growth and, within it, economic cyclicality from the Middle Ages to the end of the nineteenth century made an outstanding impression on the participants (Zimányi, 2001, 40–41). Merely the act of taking part in a conference was under the watchful eye the intelligence services. When István Varga, working at the Institute of History as a bibliographer and reporting to the intelligence services under the code name "Vili," wrote his informer's report, he found it worthwhile to comment on Vera Zimányi's attitude toward quantification methods. "She works quite enthusiastically on something that has been gaining traction internationally over the past decade, namely, deploying quantification methods to bring the processes of the past closer to the people of today."[1]

The 1960s was also the decade when Ford Foundation grants became available once more in the United States. The openly declared mission of the Ford Foundation was not limited to stimulating dialogue in the field of scientific research but also included the promotion of the American way of life (Somlai, 2010, 252). From a Hungarian perspective, the grant program had the advantage that, being managed by a foundation, it did not demand "reciprocity and equal commitment" from both sides. However, this also meant that while the Hungarian government had the right to propose beneficiaries, the foundation kept its own list of candidates for the grant. The foundation made efforts during its negotiations with the Hungarian government to have those candidates included in the program. As a rule, 25 Hungarian researchers had the opportunity to travel to the United States every year, each spending a full year there; nearly half of these grants were reserved for researchers representing the

[1] Állambiztonsági Szolgálatok Történeti Levéltára, III/III-4-a. 38 083 "Vili" M-dosszié, H-24 991. p. 24. Bp. 24 November 1970.

social sciences (Somlai, 2009, 290–292; 2021, 109–129). There was no written contract. The participants had been recruited year to year through the negotiations. György Ránki was one of the first researchers to receive the grant, with Iván T. Berend traveling the following year, in 1966 (Berend, 1997, 168). György Ránki (1930–1988), who, just like Iván T. Berend (*1929), was originally a student of Zsigmond Pál Pach, left behind an equally significant legacy in such varied areas as economic, social, political, and diplomatic history. In his internal report coming back from the Stockholm congress in 1960, Ránki made a comment that was an accurate reflection on how the younger generation thought about the new opportunities ushered in by the change of international scientific relations.[2] The long-term stays in the United States attracted the attention of the Hungarian intelligence services, so much so that they made an attempt to take charge of the process. Eventually, the Central Committee of the Party intervened to convince the intelligence services to give up on their plan to recruit secret agents from among the grant beneficiaries (Ungváry, 2006).

By the end of the 1960s, traveling to West Germany for research purposes became commonly available thanks to the Humboldt Foundation, the Friedrich Ebert Foundation, and the Goethe Institut. In 1969, 62 persons received three-to-twelve-month stipends to travel to and work in West Germany (Somlai, 2009, 313). Of course, with the gates open, there was no way to stop social scientists from considering personal invitations that came their way. When agricultural historian Péter Gunst (1934–2005), for example, received an invitation to spend a semester at the Department of Agricultural History of the Faculty of Agricultural Sciences of the University of Hohenheim, he wondered whether the invitation, which he certainly considered useful in academic terms, "had any strings attached as far as the authorities sending out the invitation were concerned." In order to prevent being approached by the foreign secret service—a scenario he believed was plausible—he approached the Hungarian Ministry of the Interior, volunteering his services as an informer.[3]

[2] Budapesti Corvinus Egyetem Levéltár, 226/a, Box 1, Folder 2. Ránki, "*Report on the 1960 Stockholm Historical Congress*," 19 September 1960.

[3] Állambiztonsági Szolgálatok Történeti Levéltára, 30 220. p. 4. "Takács Imre" 26 January 1967.

A team of two Hungarian authors would have hardly been able to break into the book market of the English-speaking world with two books on economic history had it not been for their successful participation in international conferences and for the long-term foreign grants that allowed them to travel and do research abroad (Berend & Ránki, 1974a, 1974b). In his memoirs, Iván T. Berend gives an account of the circumstances under which these two books were written. The request for the first book came right after the economic history conference that took place in Munich in 1965. Berend and Ránki's comment at the Munich conference piqued the attention of an American professor, Shephard Clough, professor of economic history at Columbia University, one of the founders of the American Economic History Association and of the International Economics History Association (hereafter the IEHA), as well as of the *Journal of Economic History*. He visited Budapest after attending the Munich conference thanks to the Pach's invitation (Clough, 1981, 237). Talking over a cup of coffee at the terrace of Café Gerbaud, he asked Berend and Ránki to write a book for the prestigious Columbia University Press about the economic history of East and Central Europe in the nineteenth and twentieth centuries (Berend, 1997, 153). What a scene of unadulterated bourgeois idyll smack in the middle of the Kádár regime's first period of softening!

While we are aware of several papers published in leading economic history journals, we have no knowledge of any other case in which Western publishers approached Hungarian authors as a result of which an actual book was eventually published. The fact that the two authors had spent one year each processing international literature as Ford grant beneficiaries must have played a major role in the fact that their book, *Economic Development in East-Central Europe*, was published successfully. And it still was not easy. The Hungarian version of the book was published in 1969; the American publisher received the English translation in the middle of 1970. As the preface of the 1974 issue lets us know, "Americanizing" the "Hungarian English" of the book took four years (and as one of the reviewers points out, that work was not always successful) (Berend & Ránki, 1974a, VIII; Eddie, 1975, 542). Another factor in favor of publishing the book was that the foreword was written by István Deák (1926–2023), who had lived in the United States as an emigrant since 1948, first studying and later teaching at Columbia University as a professor of history. Deák was an enthusiastic supporter of modern Hungarian history writing. It is also from Berend's memoirs that we know how the second book was born: at the time of the 1968 conference in Bloomington, University of Exeter Professor Walter Minchinton approached the authors to write a review of the hundred years of Hungarian economic history to be published as the first volume of a series just being launched (Berend, 1997, 152).

Both books received accolades but were also met with their fair share of criticism. As John Lampe, an American researcher of the Balkans, wrote: "Here are two books in English by Hungary's leading economic historians, plainly written from Marxist first principles and yet essentially free of the stock phrases, schematic analysis and ignorance of Western sources that is often assumed to characterize even the more recent scholarship from Eastern Europe" (Lampe, 1975, 127–128). Lampe saw the financial capital thesis and the exaggerated emphasis on the role of the banks as the weakest links of the concept; at the same time, he seemed to be only a little surprised by the fact that the figure of the entrepreneur had no place in Marxist economic history. Comparison of the Hungarian and the English texts reveals additional peculiarities, especially its theoretical grounding, which recalls Rostow's stages of growth theory (McCagg, 1975, 858). However, even the most critical reviewer acknowledged that the authors' work bridged a long-standing gap, (Palairet, 1975) and that what had been missing both in the East and in the West was an economic history discussing a larger region from a perspective capable of rising above the national one.

Besides the Institute of History at the Academy, another center of economic history research was the Department of Economic History of Karl Marx University of Economic Sciences. The department was headed by Pach until 1966 (he was also rector 1963 through 1966), followed by Berend from 1967. Most faculty members at the department were on the payroll of the university, but some of the researchers held a position at the Academy. Hence, it was not uncommon for a foreign researcher to arrive at the Institute of History as a guest of the Academy and then, as part of his or her program there, to visit the department as an off-site research center financed by the Academy. During the years between 1974 and 1979, 21 foreign historians visited the economic history department, including Peter Mathias (Oxford), Herman Van der Wee (Leuven), Michael Moissey Postan (Cambridge), Scott M. Eddie (Toronto), Thomas Kuczynski (Berlin), Abel Poitrineau (Clermont-Ferrand), Shingo Minamizuka (Chiba), Alec Nove (Glasgow), Jürgen Kocka (Bielefeld), Osamu Ieda (Tokyo), Anatoly

A. Uspensky (Moscow), Alan Milward (Manchester), Ulla Wikander (Uppsala), Helga Nussbaum (Berlin), Bo Gustafsson (Uppsala), and a group of six economic historians from Poland: Irena Kostrowicka, Zbigniew Landau, Janusz Kaliński, Wojciech Roszkowski, Stefan Wrzosek, and Zofia Gródek.

Mapping out this network of academic relationships suggests British dominance (Oxford, Cambridge, Glasgow, Manchester). There is also a link to the leading figures of the IEHA (Postan was one of the founders, while Mathias was the incumbent president of IEHA between 1974 and 1978). In this sense, the network had its own hierarchy. The second position was held by the Germans (Berlin, Bielefeld) followed by the Swedish (Uppsala) and the Japanese. We can also take a look at our imaginary map to see which centers are missing: Paris and Vienna. Stepping away from this hierarchical network of connections, we can also observe the emergence of Budapest as a research hub—a place where scholars *could* come to do research and where it was *worth* coming and doing research. When Hungarian grant opportunities started to become available, the interest they generated was certainly due to factors such as the Hungarian revolution of 1956 or the reforms of 1968, but the leftist student movements in the West also nourished an interest in the renovative capacity of the socialist system.

This was also the period when a different system of relationships was emerging around the department, this time focused on a theme that was gaining prominence within the profile of the department: East European research. The Polish initiative was launched in autumn 1974 and mostly covered post-1918 topics. The Hungarian response was a push to expand the research horizon in terms of both time and themes, first and foremost toward the non-economic factors of progress (state intervention, social behavior, mentality).[4] As the Hungarian department head put it in his 1975 report: "This year, we established closer contact with the Economic History Department of the Main School of Planning and Statistics. ... This working relationship may prove highly fruitful in the context of East European economic history research."[5] Over the course of the years, the idea germinated into a series of East Central European

[4] Author's private collection, 18 November 1974.

[5] Budapesti Corvinus Egyetem Levéltára, 316, Box 20. Berend, "*Report*," 13 September 1975.

seminars. The series was launched in Budapest and continued in Warsaw and later in Prague. Accordingly, the themes of the seminars covered an appropriately wide range of topics with a degree of asymmetry. During the 1975 seminar in Budapest, the reports focused on Polish society and on East European populations and societies between the two world wars. The papers put up for discussion during the 1977 seminar in Warsaw focused on the role of foreign capital between the two wars (Polish, Czechoslovak, Hungarian capital import and German capital export). By the time the seminar was organized in Prague in 1978, the focus had shifted to the structural transformation of the economies of the European socialist countries after 1945. The perspective had become more international, and the people's democracies took center stage as the protagonist group of the story.

One might reasonably expect that such established and institutionally supported contacts would have transformed into closer cooperation, especially recalling social network theoretician Mark Granovetter's famous paper "The strength of weak ties" on networks based on institutional rather than strictly personal contacts (Granovetter, 1973). Taking a closer look at this 1970s series of seminars makes one think of the exact opposite: the weakness of weak ties. The Czechoslovaks had already attended the seminar in Warsaw, and by the time Prague hosted the event, the universities of Bratislava and Berlin were also on board. As far as I recollect, the idea of inviting the Soviets was floated already in Prague, although both the Polish and the Hungarians refused to consider the suggestion, and by the 1980s the entire initiative died away. This is notable because by that time the idea of project-based research cooperation, which had gained traction all over the world, was beginning to take root in Hungary. Apparently, the East Central European perspective failed to provide a fertile enough research paradigm for the historiographies of countries heavily accustomed to writing history from a national perspective.

At the time, the Polish–Hungarian "axis" had no Visegrád overtones; however, the parties that tried to join forces were two of the most highly advanced national centers of economic history research in Eastern Europe. What we see was a certain East Central European attempt to create a coalition for coordinated self-defense. The Budapest Department of Economic History wielded the most clout by the 1980s, which implied a favorable shift in the academic pecking order in terms of allocating the conference themes within the IEHA. However, these meetings never germinated into any joint research projects.

On Open (Battle)fields

Historical literature on the Cold War has by now shaken off the shackles of approaches focusing narrowly on political history, and, just as in other fields of study, the analysis of transnational cultural interrelations now plays an increasing role (Gienow-Hecht, 2000, 2010). As far as the historiographical review of our own discipline is concerned, the first important steps have already been taken in the context of East–West relations (Berg, 2015). Viewed from this perspective, the extent to which politics and ideology determined academic contacts—possibly to the detriment of the interests of the profession—is certainly one of the decisive factors. On the other hand, during the period under review, this determination clearly depended on the institutions that made the decisions on financing academic activities, as well as on whether there existed organizations (foundations, academies) that enjoyed at least relative autonomy from political power and that were available as potential sources of funding for academic activities driven by the intrinsic rules of the profession rather than politics.

One of the most important international organizations in the field of history writing, the Comité International des Sciences Historique (hereafter the CISH), made laudable efforts to relaunch the professional dialogue among historians of different countries after the war. There were many open questions to resolve: the continuation and renewal of the membership of the losing countries, the rights of representation of the legitimate leaders of the national member organizations, and so on. A change of the prewar old guard at the helm of the CISH was also on the agenda. When in April 1948 the General Assembly convened in Paris in honor of the one hundredth anniversary of the revolution of 1848, the Soviet Union refrained from participation, but Hungary was represented by two senior officials of the Hungarian Historical Society (hereafter the Society), President Ferenc Eckhart (1885–1957), and Acting President Domokos Kosáry (1913–2007) (Erdmann, 1987, 273).

As a new board member of the Society, Pach expressed his opinion on the CISH while being interrogated on Kosáry by the State Protection Authority in 1950: CISH is "the association of anti-Marxist Western

European historians."[6] The Faculty of Arts of Pázmány Péter University was invited to the CISH congress in the summer of 1949, but the decision was escalated to Magyar Tudományos Tanács [the Hungarian Council for Sciences], the agency in charge of the Sovietization of academic life in Hungary.[7] Mrs. Erzsébet Gerő (née Fazekas, 1900–1967), who was the secretary of the Section of Social Sciences of the Hungarian Council for Sciences and the wife of Ernő Gerő, number two of the Muscovite troika leading the Party during the Stalin era, knew well from where the wind was blowing: "I also have no knowledge of whether the Soviet Union deems participation at that congress necessary, although that is exactly what everything depends on."[8]

When the first postwar International Congress of Historical Studies convened in Paris in 1950, the Soviet Union was no longer the only country not to delegate historians. The most concentrated international connections for the historical sciences as an arena were produced by the congresses. The return of the Soviet Union and other people's democracies to the Rome congress in 1955 created the possibility for East–West dialogue: the five-member Hungarian delegation was led by Erzsébet Andics (1902–1986) (Andics, 1955, 855). She played a leading role in the Soviet-type "Gleichschaltung" of the Society (and Hungarian historiography at large) (Kövér, 2016; Romsics, 2011, 356–377). None of the delegates was an economic historian and only a single Hungarian comment—Andics's account of the ten years of Hungarian history writing since 1945—became part of the official program. None of the papers presented by Hungarians were about economic history. In turn, the Hungarian delegation tried to take every opportunity to make oral interventions and managed to make their voices heard on as many as ten occasions. Interventions of this type were mostly possible in the context of discussions around issues related to economic and social history and the history of social and political movements, topics that offered possibilities for cooperating with not only East European but also West European Marxists.

[6] Állambiztonsági Szolgálatok Történeti Levéltára, O 11182 "Kosáry Domokos," 27 June 1950.

[7] Eötvös Loránd Tudományegyetem Levéltár, 8.a. 51. pp. 184–185. "*Minutes of the Faculty Council*," 14 July 1949.

[8] Magyar Tudományos Akadémia Levéltár, 2. MTT, Box 1, Folder 11. Gerőné Fazekas, Erzsébet: "*Memorandum*," Bp. 20 September 1949.

The next congress coming up in Stockholm, a neutral country, in 1960 was to take place during a period marked by the gradual organizational secession of economic historians and the organization of the first international conference on economic history (Berg, 2015). In his report on the Stockholm congress, Braudel, one of the IEHA's principal organizers, welcomed the presence of historians representing the Soviet Union and the people's democracies, as he believed that Marxism as a global interpretation of society and history exemplified the application of "a long-term model" (Braudel, 1961, 498–499). The members of the Eastern camp made concerted efforts to prepare for being put to their first international test: they held preliminary discussions in Prague and Leipzig. From the first one (called "informational consultation") we have the Russian language protocols, for the second one we have only a short oral quotation by Ránki, at that time secretary of both the Hungarian Committee for Historical Studies and the Hungarian Stockholm delegation.[9]

In line with a 1957 resolution of the Preparatory Committee for the Stockholm congress, only three comments were allocated to Hungarian delegates (Director of the Institute of History, Erik Molnár 1894–1966, Makkai, and Pach). These comments had been discussed previously in Budapest (July 11, 1960).[10] Moreover, a two-volume foreign language compilation of Hungarian papers was published on the occasion of the international congress, under the title *Études Historique*. These volumes (fitting to the immediate post-1956 period) laid more of an emphasis on the "popular front" approach by allowing the older, non-Marxist generation of Hungarian historians to contribute, while both the older and the newer generations weighed in more heavily on topics falling within the realm of economic and social history. The Hungarian delegation had brought 150 copies with them to Stockholm and reported "having run practically out of stock within just a day" (Ránki, 1960, 760). Not one of the non-Marxist contributors of the volume took part personally in the congress.

Pach's communication at the congress (Kövér, 2020) contributed to the debate on a report by two fellow Eastern Bloc historians, Arnošt

[9] Magyar Tudományos Akadémia Levéltár 224. "Papers of György Ránki," "*Protokol, Prague 18 September 1959*"; Budapesti Corvinus Egyetem Levéltár, 226/a. Box 1. Folder 2. "*Minutes of the 11 July 1960*".

[10] Budapesti Corvinus Egyetem Levéltára, 226/a, Box 1, Folder 2. "Papers of Zsigmond Pál Pach."

Klíma of Prague and Josef Macůrek of Brno, under the title "On the Issue of the Transition from Feudalism to Capitalism in Central Europe during the 16th and 17th Centuries." Pach's central claim was that: "... an important part of these antecedents is the divergence of Hungary's own agricultural development from that of Western Europe, a process that started at the end of the fifteenth century and came into full bloom during the first half of the seventeenth century" (Pach, 1961b, 1). The Hungarian historians participating in the debate of these theses in Budapest had not dared or wanted to engage in a confrontation. The doubts they formulated were not so much about Pach's concept of divergence but rather about the concept of the transition and how the concept of divergence relates to it. After his study trip to Moscow in the autumn of 1954, Pach revised his earlier position on the primitive accumulation of capital, holding a series of consultations with many of the most prominent Soviet experts on the topic, including senior research fellow of the Soviet Academy, Vladimir M. Lavrovsky (Pach, 1955). According to the Hungarian report on the Stockholm debate, the position laid out by Pach in his comment—"which shed additional light on the development of Eastern Europe facilitating deeper understanding"—found acceptance mostly with the Soviet commentators (primarily V. M. Lavrovsky).[11]

The first main theme of the economic history conference was industrialization, and the debate centered around W. W. Rostow's 1959 theory of the stages of economic growth. The language used by this breed of economic historians was all too foreign to the Hungarian conference participants, and neither of them intervened. However, when commenting on the second day, which was dedicated to agricultural history, they found more positive aspects to highlight the strength of Marxist economic history.[12] Pach commented on a report delivered by Munich professor Friedrich Lütge under the title "A Comparative Study of Large-Scale Agricultural Plants Since the End of the Middle Ages." Pach criticized Lütge for not considering the alternative, namely, the possibility of peasant holdings developing into large-scale holdings (a comment probably not without political overtones during the period of agricultural collectivization) (Pach, 1961a, 226–235).

[11] Budapesti Corvinus Egyetem Levéltára, 226/a, Box 7, Folder 27.

[12] Budapesti Corvinus Egyetem Levéltára, 226/b, Box 2. Papers of Zsigmond Pál Pach, October 1960.

Fourteen Hungarian delegates traveled to participate in the congress (most enjoyed the financial support of either the Academy or of their university; three delegates received state funding to cover half of their costs and financed the rest privately). Five delegates (Pach, Júlia Puskás, 1929–2005, Ránki, Berend, and legal historian Pál Horváth, 1928–2021) participated in the economic history conference[13] (Berend, 2018, 186).

Because of the Berlin crisis, the French government refused to grant entry visas to the East German economic historians for the second conference in Aix-en-Provence in 1962 (Berg, 2015, 62–63; Pach, 1963, 256). As early as the summer of 1962, Pach received a reassuring message from Jürgen Kuczynski on this issue: "...die Sowietfreunde auf jeden Fall fahren werden."[14] So the Hungarians traveled also.

Seven Hungarian historians registered to participate in the Munich conference. Carrying on with the topics successfully introduced in Stockholm, Pach—as rapporteur—joined W. Kula's section on the agricultural history of the modern period (!). One of the delegates commenting on his report was Makkai. Makkai contributed important new sources to the study of this field (not always in support of Pach's concept). In the section dedicated to issues surrounding the accumulation of capital, another Hungarian participant, Ránki, presented his views on the basis of the more recent achievements of Hungarian Marxist industrial history (Berend & Ránki, 1968; Pach, 1965).

Berend and Ránki, allocated to Section 9, had the opportunity to comment on British Marxist Eric Hobsbawm's presentation on the structural aspects of the emergence of the working class. They scrutinized the issue from the perspective of the East European specificities of the Industrial Revolution. The authors found a way to comment on the Hobsbawm-Hartwell debate about the effect of the industrial revolution on the living standards of the working class. While the first part of the presentation was dense with interpretations of quotes from Marx and Lenin and references to East European Marxist historians, the last part, in which the authors discussed the Hobsbawm-Hartwell debate, avoided taking a biased position: "Eastern-European data furnish no support or

[13] Budapesti Corvinus Egyetem Levéltára, 226/a, Box 1, Folder 2. Ránki, "*Report on the 1960 Stockholm Historical Congress*," 19 September 1960.

[14] Budapesti Corvinus Egyetem Levéltára, 226/a Box 10. Folder 50. Kuczynski to Pach, 21 June 1962.

confirmation of any simple viewpoint in a debate on British or Western-European conditions" (Berend & Ránki, 1968, 210). Even Hobsbawm himself commented positively on this contribution in his closing remarks (Pach, 1966, 21).

The Munich conference had special personal significance for the Hungarian delegation in that, in acknowledgment of his achievements, Pach was offered a post in the Executive Committee, now officially constituted. Four out of the ten posts were assigned to scholars from the socialist countries (Pach, 1966). The fact that the decision on the host cities of the following two conferences (Bloomington–Leningrad) was adopted that much in advance speaks volumes of the diplomatic clashes and horse trading that must have taken place behind the scenes before a balanced enough solution acceptable to all the great powers was eventually hashed out.

Mostly because of financial difficulties, significantly fewer Hungarian deputies could travel to Bloomington in 1968 than to Munich despite the fact that the organizers had made every effort to secure grants for the foreign delegates, especially for those from Eastern Europe (the total number of participants was less than three hundred, of which only about 80 researchers arrived from Europe). Six Hungarian historians attended the conference. All of them were faculty members of Hungary's two leading workplaces in the field of economic history research: Pach, Berend, Ránki, Makkai, Katus, and Ágnes R. Várkonyi (1928–2014), the latter that time secretary of the Hungarian Committee for Historical Studies (Pach, 1969, 67). In Section I, which carried on work on the wider theme of the emergence of capitalism, Pach organized the third session, and he was also the presenter of the report. Originally, the session centered around the favorable and unfavorable conditions for the development of capitalism; however, Pach's presentation shifted the focus to the fifteenth-seventeenth century rearrangement of international trade routes (Pach, 1968). Pach managed to recruit an international pool of highly qualified historians from the British Hartwell through the Polish Topolski all the way to the Russian Kovalchenko (although, despite the best efforts of the American organizers, the East German Jürgen Kuczynski was not granted a visa).[15]

[15] Magyar Tudományos Akadémia Levéltár, 224. "*The correspondence of Zsigmond Pál Pach*," Kuczynski to Pach, 18 October 1968.

Section II, chaired by Peter Mathias, focused on changes in the structure of industry throughout the twentieth century. Besides a report by the Czechoslovak Alice Teichova, another Eastern European presenter was Berend, one of whose commentators was Ránki. In his trip report, Pach drew the reassuring conclusion that "most of our Western counterparts place great emphasis on maintaining and improving our relations, and indeed it is only in this spirit that the IEHA can envision its continued operation going forward"[16] (Pach, 1968).

As can be expected, the American apostles of the new economic history (D. C. North, R. Fogel) took center stage during the Bloomington conference of the IEHA, and there was no way not to acknowledge the methodological innovations they put forward. Ever since the French–Hungarian seminars of economic historians, the Hungarians had been aware that quantification had been gaining terrain. "... to our great satisfaction, we were able to establish that the sort of gratuitous modernity, the drive to mathematize everything at all cost, and the contrived application of formulas to historical processes—in other words, the methodological extremes pushed way beyond what is healthy and fruitful—barely gained any traction during the congress... [and] did not eventually led to the burgeoning of the methods we considered overused" (Pach, 1969, 80–81). Seeing Pach adopt this attitude toward quantification is intriguing especially in light of the fact that he had quite a few family members and close friends who were prominent mathematicians.

In Leningrad in 1970 Hungary was represented by a larger-than-ever delegation of twenty researchers among the 1,360 registered conference participants (Berend, 1971). The Leningrad conference placed much greater emphasis on theoretical and methodological issues than any previous congresses. Section IX probably took the lead in this, as it was focused around a report by Hungarian authors Berend and Ránki entitled "*Universal Economic History.*" The authors formulated their methodological concept in the spirit of regional comparative economic history; at that time, they were already working full steam on their book on economic development in East Central Europe. Another session of this section was dedicated to quantification. Here, the French rapporteur was followed by sixteen comments and almost as many oral interventions; one of the most

[16] Magyar Tudományos Akadémia Levéltár, 224. "*The papers of Zsigmond Pál Pach,*" Box 2/II Folder 27/c. Pach, "*Report on my Trip to the United States on the Occasion of the 4th Congress of the International Economic History Association,*" 8 November 1968.

important takeaways being that the application of computer technology in the field of historical research was gaining traction not only in the West but also in several Soviet centers of economic history research. The Hungarian conference report once again underlined "the deficiencies of the new methods" (Berend, 1971, 256–257). The fact that "the comparative method has continued to gain terrain" during the congress was met with much more positive reception, almost to the degree of sensing "a breakthrough in attitude" (cf. Baráth, 2021).

The interest in international economic history conferences seems to have declined after Leningrad. Proper conferences were only convoked once every four years, although the Executive Committee held several sessions in between (Prato, 1971; Bellagio, 1972; Copenhagen, 1973), providing new Danish president Kristof Glamann with an opportunity to renew the structure in which the conferences were organized (Glamann, 2002, 88–89). The "Section A" modules also saw a thematic renewal: the topics introduced in 1974 included the environment, urbanization, and regional economic development inequalities. Hungarian journals published fewer triumphant conference reports during this period than in the 1960s, despite the fact that the Hungarian delegations participating in Copenhagen and in Edinburgh were just as active as any of the previous Hungarian delegations. A Hungarian scholar with the opportunity to present a "Section A" report was Iván T. Berend in the entrepreneurial history section analyzing investment strategies in the nineteenth and twentieth centuries. Here, the concept was based on Alfred D. Chandler's theory, which revolutionized business history. The book was published on the occasion of the conference (Daems & van der Wee, 1974). Berend was the only East European author featured in the book with a chapter on "Investment Strategy in East-Central Europe" (op. cit., 170–197). During the conference, Alexander Gerschenkron commented on Berend's paper, questioning, among other things, his calculations on capital import: "It is my impression that Mr. Berend on the whole tends to exaggerate the quantitative importance of foreign investment" (Gerschenkron, 1978, 71).

A closer look behind the scenes reveals new aspects to the role played by the Hungarians. We have already referred to the fact that the relations between the East and the West had always implied some degree of internal consultation among the countries of Eastern Europe. As Pach put it in one of his internal reports, "… it is safe to say that a form of socialist cooperation has fully emerged within the Association. The representatives of the socialist countries... meet separately prior to the congresses to discuss how best to prepare for the expected debates both in terms of content and contributors. We held such a preparatory meeting in Budapest both before the Leningrad congress of 1970 and prior to the Copenhagen congress of 1974, enjoying the full support of the competent unit of the Hungarian Academy of Sciences."[17]

The preliminary meeting leading up to the Leningrad congress was held June 13–15, 1970, while the one before the Copenhagen congress took place between July 11–12, 1974. The representatives of the Czechoslovak and Yugoslav academies did not participate in the preliminary meeting organized ahead of Leningrad. The Leningrad conference was significant not only because it was the first time the event was hosted by a socialist country, but also because it "offered the widest possible international forum for the confrontation of the achievements and methods of Marxist and bourgeois economic history." Accordingly, in the preliminary meeting leading up to the congress, "the socialist delegations are making especially careful preparations for the sessions of Section I (History of Economic Thought) and Section II (Economic Planning), where the harshest debates may be expected to take place."[18] By 1974, during his opening address at the beginning of the preliminary meeting—by then officially termed "coordination meeting"—Pach said they were preparing for "holding our ground and consolidating our achievements during the Copenhagen conference." The Czechoslovak delegate canceled his participation because of health-related reasons; the Bulgarians and the Yugoslavs did not send delegates.[19]

In the meantime, significant changes took place on the global political scene: the signing of the Helsinki Accords, a declaration on the spirit of détente in international relations, also had a major effect on the conferences. The reception of the toast delivered by Giscard d'Estaing during his visit in Moscow in the autumn of 1975 accurately reflected

[17] Magyar Tudományos Akadémia Levéltár. 224. "*The papers of Zsigmond Pál Pach,*" Folder 30. Pach, "*Report on the Activities Performed Between 1970 and 1975 in the International Economic History Association,*" 31 October 1975.

[18] Magyar Tudományos Akadémia Levéltár, 224. "*The papers of Zsigmond Pál Pach,*" Folder 59. Pach, "*Report,*" 27 June 1970.

[19] Magyar Tudományos Akadémia Levéltár, 224. "*The papers of Zsigmond Pál Pach,*" Folder 59. Pach, "*Opening Address,*" 12 July 1974, and Pach, Zsigmond Pál to President of the National Committee of Hungarian Historians Dezső Nemes, 15 July 1974.

the diverging interpretations of the détente between the West and the East. While talking about the so-called third basket Giscard expressed his desire for an "ideological détente." Brezhnev, by contrast, held that "the intensification of the ideological struggle (third basket) is expected to become a lasting trend." This was an indication that the science policy of the Eastern Bloc held on to ideological confrontation as one of its fundamental ambitions.

The economic changes taking place throughout the 1970s also had an impact on the themes taken up by the economic history conferences. By the time of the 1978 Edinburgh conference, the end of the golden era of growth became manifest: there was only one section dedicated to economic growth, although that topic had always been a central issue since 1960 (Rév, 1979, 181). The themes that were brought into the forefront instead included "Natural Resources and Economic Development and Regional and International Inequalities," with two Hungarian presentations being delivered by Ránki and Éva Ehrlich (Rév, 1979, 183; van der Wee, 1993, 101). The appeal of the comparative approach drove a search for ever-more global issues to tackle. In the meantime, the theoretical background also shifted toward theories of underdevelopment.

The internal report of 1975 on "socialist cooperation" also included the first unambiguous reference to Budapest being a favorable choice as the host city of an upcoming conference: the Soviet representative of the IEHA's Executive Committee proposed that the congress after 1978 "should take place in Budapest," with Pach elected as president of the IEHA.[20] Although during the Leuven meeting of the Executive Committee in 1977 most members preferred Switzerland, they failed to convince the Soviets, who stubbornly insisted on their own stance.[21] Eventually, following the joint proposal of the British president of the IEHA and of the Soviet member of the presidency, the Edinburgh General Assembly of 1978 elected Pach as president of the IEHA, and Budapest received the green light to organize the eighth international economic history conference in 1982 (Bern becoming the host city in 1986).

[20] Magyar Tudományos Akadémia Levéltár, 224. "*The papers of Zsigmond Pál Pach,*" Folder 30. Pach, "*Report on the Activities Performed Between 1970 and 1975 in the International Economic History Association,*" 31 October 1975.

[21] Magyar Tudományos Akadémia Levéltár, 224. TI "*Miscellaneous Papers,*" Box 9. General Secretary P. Jeannin to Pach, 22 December 1977.

Of course, organizing the conference in Budapest required a declaration of intent from the Academy to host the event. As soon as the decision was adopted in Edinburgh, the promoters of the conference had to lobby the various fora of political leadership in an effort to convince them that such an event would serve the best interests of the country as a whole. The first politician who had to be won over was György Aczél, the chief pundit of cultural and ideological affairs within the Party, whom Pach managed to meet on February 28, 1980.[22] Next came the various fora of the Party state apparatus such as the Coordination Committee for the Social Sciences (February 13, 1980), the Science Policy Committee (April 12, 1980), and the Agitation and Propaganda Committee of the Central Committee of the Party (July 1, 1980). As far as financing the conference was concerned, the leadership of the Academy itself took charge of the negotiations with the Ministry of Finance. The ministry "acknowledged" "the utilization of 500 thousand forints from the budgetary surplus of the Hungarian Academy of Sciences as a form of budgetary support" to cover the costs of the congress.[23] The head of the Local Organizing Committee, Ránki, knew what was to be expected: the only event hosted by an Eastern Bloc country—and hence the only example Hungary could follow—was financed from the resources of the Academy of the host country (in other words, the government budget).

After the usual preparatory sessions (Neu-Brandenburg, East Germany, 1979; Altdorf, West Germany, 1980), the Executive Committee convened in Visegrád, Hungary, in 1981 to finalize the conference program. The themes for the Sections A and B were selected through the standard procedure by 1979.[24] The coordination meeting of the socialist bloc countries leading up to the Budapest conference was held in Moscow and provided an opportunity to present a progress report on the conference preparations. During the coordination meeting, the discussions revolved around "the measures to be taken to increase the impact of the researchers

[22] Magyar Tudományos Akadémia Levéltár, 224. TI "*Miscellaneous Papers*," Box 9. Pach, Zsigmond Pál to Aczél, György, 19 May 1980.

[23] Magyar Tudományos Akadémia Levéltár, 224. TI "*Miscellaneous Papers*," Box 9. "*Proposal in the Matter of the Organisation of the 8th International Economic History Congress in Budapest in 1982*," 22 January 1980; Deputy Minister of Finance Lajos Kállai to General Secretary of the Hungarian Academy of Sciences István Láng, 23 August 1981.

[24] Pach, "*Notes on the 20–21 April 1979 Meeting of the Executive Committee of the International Economic History Association in Neu-Brandenburg*." Ibid.

of the socialist countries" in history congresses in general, as well as the preparations for the economic history conference.[25] Pach presented a report on the latter in Russian. Pach mentioned the fact that it was his own suggestion during the Visegrád Executive Committee meeting that the number of Section C modules be increased from 31 to 36 after he had approached the Bulgarian, Romanian, and Yugoslav National Committees to propose themes they wished to curate as co-organizers. The amendments were approved by the Executive Committee; hence, ten out of the 36 Section C modules involved historians from the socialist countries. The socialist countries took special care to ensure the existing proportionality within the Executive Committee (two members representing the Soviet Union and one Poland, Czechoslovakia, East Germany, and Hungary each). The whole-day meeting was followed by a "dinner of comrades ["товарищеский ужин," "tovarishchesky uzhin"] at 18.30."[26]

Held August 16–20, 1982, the Budapest conference took place with the participation of a record number of Hungarian delegates (104 registered Hungarian researchers). The official report on the congress reports on even more, 155 Hungarian participants (The Eighth, 1983, 211). Of course, we have no data on how many were actually present, but this was the first opportunity for the younger generation of Hungarian economic and social historians to experience the atmosphere of such an international conference. In his welcome speech on behalf of the Local Organising Committee, Ránki—remembering as a lucky permanent participant of all congresses from the beginning—put Hungarian economic history on the European scientific map in between "the French economic historians with the circle of Annales and the only economic history institute of the East European part of the world," the economic historians of GDR, with whom "we are pleased to have a permanent series of meetings."[27]

Three Section A modules focused on the traditional main themes (large-scale holdings versus small holdings, protoindustrialization, technological change). Makkai, who had shown a remarkable ability to renew himself ever since 1960, presented the report on large-scale versus small

[25] Magyar Tudományos Akadémia Levéltár, 224. "*The papers of Zsigmond Pál Pach,*" Folder 638. Tikhvinsky, S. L. to Nemes, Dezső, 24 April 1981.

[26] Magyar Tudományos Akadémia Levéltár, 224. "*The papers of Zsigmond Pál Pach,*" Folder 638. "*Report by Zsigmond Pál Pach, Moscow, 26 June 1981.*"

[27] Magyar Tudományos Akadémia Levéltár 224. MTA TI "*Papers of György Ránki*" Box 15. The drafts of Ránki;s welcome speech.

holdings. In the section on protoindustrialization, F. Mendels presented a report that failed to arouse any interest in Hungary even though it was published in Hungarian on the occasion of the congress.

Taking a look at the conference program, it is obvious that quantitative research methods and the study of long-run trends and cyclical fluctuations gained prevalence. The discipline continued to widen its horizon toward social history, family history, the history of women, and interdisciplinarity. In the section organized by Estonian J. Kahk on quantitative economic and social history methodologies, one of the presenters—in addition to Robert W. Fogel—was György Granasztói, who, along with his French co-author, focused on factor analysis and the longue durée approach to the study of history (Demonet-Granasztói, 1982). In Section B, which was organized by Leslie Hannah under the title "From Family Firm to Professional Management," Péter Hanák gave an inspiring presentation that reviewed the results of an entrepreneurial history project involving a team of young Hungarian researchers. The only confusion that arose during the conference had to do with the innovative open forum event during which the organizers (led by Swiss Vice President Jean-François Bergier) set the tone for the evening session by presenting a documentary film on the Russian famine of 1921 in Saratov. Understandably, the economic historians of the Third World found the topic especially interesting, but the Soviet comrades found the choice of topic unpalatable.

During the Budapest conference, the General Assembly elected Bergier president of the association, despite the fact that Fernand Braudel tried to prevent his appointment. As Pach was promoted to honorary president, the position he left vacant on the Executive Committee was filled by Iván T. Berend.

The following conference was—once again—organized in a non-allied country, Switzerland, with the city of Bern acting as the host. The end of the Cold War was drawing near and a new era, hallmarked by the Soviet Union's perestroika movement, was about to begin. This is not to say that this conference was any less ridden with ideological clashes. Just one example should suffice—the main section debate organized by Iván T. Berend and Knut Borchardt under the title "The Crisis of the 1930s and its Relevance to Our Own Time" (Berend & Borchardt, 1986). The Soviet V. M. Vinogradov and the Bulgarian L. Berov promoted the ideologically loaded position that "in theory, a crisis is not possible in a socialist economy; such phenomena are merely ripple effects that derive from the crisis of capitalism and propagate through the intricate relationships that link socialist countries to the world economy." This position found its critic in György Ránki, already teaching a semester in every year in Bloomington at the time, who claimed that "the failure to acknowledge the crisis may lead to the thwarting of innovation, the failure to proceed through with structural transformation, and falling behind the world economy" (Csató, 1988, 468).

As a sign of the new times, the program of the conference was supplemented with a module entitled "International Research Projects in Progress." Within this module the Russian V. Bovykin and the American R. Cameron organized a thematic block under the title "The International Banking System and Industrial Financing (1870–1914)." The initiative was born during the 1982 congress in Budapest. After IREX financed a number of visits to Moscow and Atlanta, a pre-conference was hosted by the Bellagio Center of the Rockefeller Foundation with the participation of 26 researchers (Cameron, 1985).

The Bern congress was clearly a breakthrough in that it was dominated by themes related to the modern era, a trend that came to full bloom during the Leuven congress in 1990. Undoubtedly, the single greatest innovation was the decision to institute the doctoral section (Section D), which allowed the youngest generation of congress goers to make their voices heard from the speaker's lectern (Berg, 2015, 69; van der Wee, 1993, 103). It was also the last congress whose ambitions still reflected—symbolically at least—the dwindling dominance of the traditional East–West dimension: 80 East and Central European scholars and 40 researchers from the Third World received foundation support so they could participate (van der Wee, 1993, 95). Once the Cold War was over, the focus of the congresses shifted to the North–South division. The opening toward the Third World—a move much urged by both parties—had unexpected consequences.

Conclusion

Under the circumstances of the Cold War, the small nations of East Central Europe and their economic history writings had only limited and changeable scope for action. However, in practice, partly secretly, semi-officially, and later in more institutionalized, but controlled forms (international agreements, scholarships, and so on), new dimensions for real scientific contacts and communications opened. During the 1970s, two

Hungarian economic historians were able to break into the book market of the English-speaking world. At the same time Budapest became a place where scholars came to do social science research. Concerning the main battlefields—international conferences—the relations between the East and the West had always required internal preparatory meetings among the countries of Eastern Europe (Prague, Leipzig, Budapest, Moscow). The initiative for the Budapest conference in 1982 also originated from Moscow. Even so, on the base of established international contacts and scientific embeddedness, Hungarian economic history writing was able to grow out from this "advanced post" position and fulfill a mediator function between East and West. In the circumstances of perestroika, the surrounding ice floes started gradually to thaw.

Hungarian economic history research reached its pinnacle in the 1980s, arriving at a high plateau where merely maintaining the standard required considerable effort. Organizing the economic history congress successfully (the Budapest event was praised by all participants in superlatives) was only one of the factors explaining that achievement; just as importantly, the 1980s was a decade when an entire new generation of young economic historians started their careers, most of them mentored by Ránki, who unfortunately passed away in 1988. This new generation of historians contributed to countless international projects in a whole range of research areas (agricultural history, banking history, business history, and so on) (Pogány, 1999). Typically, the teams that reached out to these young Hungarian researchers were organized by scholars of whom at least one was either based in, or had emigrated from, Eastern Europe. These projects either continued or were occasionally renewed even after the change of the political regime. However, the year 1990 marked not only the end of the Cold War and communism in Eastern Europe but also the end of the heyday of Hungarian economic history research.

Translated by Attila Török.

References

Állambiztonsági Szolgálatok Történeti Levéltára [Historical Archives of the Hungarian State Security].

Eötvös Loránd Tudományegyetem Levéltár [Eötvös Loránd University Archives].

Magyar Tudományos Akadémia Levéltár [Archives of the Hungarian Academy of Sciences].

Andics, E. (1955). A római történészkongresszus, *Századok, 89*, 835–848.

A Történettudományi Intézet Értesítője. (1950). ["*Newsletter of the Institute of History*"].

Baráth, K. (2021). *Térfoglalás és reprezentáció gazdaságtörténész módra: magyarok a Nemzetközi Gazdaságtörténeti Kongresszusokon (1960–1986)* (manuscript).

Berend, T. I. (1971). Az V. Nemzetközi Gazdaságtörténeti Kongresszus és tanulságai (Leningrád, 1970 augusztus). *Századok, 105*(2), 249–271.

Berend, T. I., & Ránki, G. (1968). Peculiarities of Industrial Progress in Eastern Europe and the Development of the Working Class. In D. E. C. Eversley (Ed.), *Troisième Conférence Internationale d'Histoire Économique / Third International Conference of Economic History*: Munich 1965 (pp. 193–213). La Haye, Mouton.

Berend, T. I., & Ránki, G. (1974a). *Economic Development in East-Central Europe in the 19th and 20th Centuries*. Columbia University Press.

Berend, T. I., & Ránki, G. (1974b). *Hungary, a Century of Economic Development*. Barnes and Noble Books.

Berend, T. I., & Borchardt, K. (Eds.). (1986). *The Impact of the Depression of the 1930s and Its Relevance for the Contemporary World*. Karl Marx University of Economics, Academy Research Center of East-Central Europe.

Berend, T. I. (1997). *A történelem – ahogyan megéltem*. Kulturtrade.

Berend, T. I. (2018). *Naplementék (történelmi korszakvázlatok)*. Éghajlat Könyvkiadó.

Berg, M. (2015). East-West Dialogue: Economic Historians, the Cold War and Détante. *Journal of Modern History, 87*(1), 36–71.

Braudel, F. (1961). Stockholm 1960. *Annales. Economies, Sociétés, Civilisation, 16*(3), 497–500.

Budapesti Corvinus Egyetem Levéltára [Archives of Corvinus University of Budapest].

Cameron, R. (1985). Introduction to the Cameron-Bovykin Project. *Business and Economic History, 14*, 17–18.

Clough, S. B. (1981). *The Life I've Lived. The Formation, Career, and Retirement of an Historian*. University Press of America.

Csató, T. (1988). Beszámoló a IX. Nemzetközi Gazdaságtörténeti Kongresszusról (Bern, 1986. augusztus 24–29.) *Századok, 122*(3), 464–471.

Czoch, G. (2015). Az Annales történetírásának magyarországi recepciója az 1960-as években. In P. Erdősi & J. Klement (Eds.), *Vis-à-Vis. Tanulmányok az Atelier fennállásának 25. évfordulójára* (pp. 111–143). Atelier Európai Társadalomtudományok és Historiográfia Tanszék Korall Társadalomtörténeti Egyesület.

Daems, H., & van der Wee, H. (Eds.). (1974). *The Rise of Managerial Capitalism*. Leuven University Press and Martinus Nijhoff.

Demonet, M., & Granasztói G. (1982). Une ville de Hongrie au milieu du XVIe siècle: Analyse factorielle et modèle social. *Annales. Economies, Sociétés, Civilisation, 37*(3), 523–551.
Eddie, M. S. (1975). *Economic Development in East-Central Europe in the 19th and 20th Centuries* by Ivan T. Berend & György Ránki. *Canadian Slavonic Papers / Revue Canadienne des Slavistes, 17*(2–3), 542–544.
Erdmann, K. D. (1987). *Die Ökumene der Historiker. Geschichte der Internationalen Historikerkongresse und des Comité International des Sciences Historique*. Vandenhoeck & Ruprecht.
Études des délégués Hongrois au Xe Congrès International des Sciences Historique. Rome, 4–11 septembre 1955 [Special number of *Acta Historica* 4. nos. 1–3. Eds. E. Andics et al.].
Etudes historique publiées par la Commission Nationale des Historiens Hongrois. I. (1960) Budapest.
Gerschenkron, A. (1978). Comments. In K. Glamann & H. van der Wee (Eds.), *Sixième Congres International d'histoire économique. Copenhague, 19–23 aout 1974. Actes/Proceedings* (pp. 70–72). Akademisk Vorlag.
Gienow-Hecht, J. C. E. (2000). Shame on US? Academics, Cultural Transfer, and the Cold War—A Cultural Review. *Diplomatic History, 24*(3), 465–494.
Gienow-Hecht, J. C. E. (2010). Culture and the Cold War in Europe. In *The Cambridge History of the Cold War. v. 1. Origins* (pp. 398–422). Cambridge University Press.
Glamann, K. (2002). SEHR & IEHA—The Early Years. *Scandinavian Economic History Review, 50*(3), 83–89.
Granovetter, M. S. (1973). The Strength of Weak Ties. *American Journal of Sociology, 78*(6), 1360–1380.
Katus, L. (2008). Meghallani a sokszólamú zenét, az emberi történelmet" Bódy Zsombor és Cieger András beszélgetése Katus Lászlóval. In Nagy, M. & Vértesi, L. (Eds.), *Sokszólamú történelem. Válogatott tanulmányok és cikkek* (pp. 423–461). PTE BTK.
Kövér, G. (2009). The Economic Achievements of the Austro-Hungarian Monarchy. Scale and Speed. In A. Gerő (Ed.), *The Austro-Hungarian Monarchy Revisited* (pp. 51–83). Social Science Monographs.
Kövér, G. (2016). Crossroads and Turns in the Hungarian Economic History. In F. Boldizzoni & P. Hudson (Eds.), *Routledge Handbook of Global Economic History* (pp. 242–257). Routledge.
Kövér, G. (2020). Kanyarodás keletre, nyitás nyugatra: Pach Zsigmond Pál pályafordulása. *Századok, 154*(6), 1255–1291.
Lampe, J. R. (1975). *Economic Development in East-Central Europe in the 19th and 20th Centuries* by Iván T. Berend and György Ránki. Columbia University Press, 1974; *Hungary, a Century of Economic Development* by Iván T.

Berend and György Ránki. Barnes and Noble Books, 1974; *Business History Review, 49*(1), 127–129.
Le Goff, J. (1968). Van-e Annales-iskola? *Századok, 102*(1–2), 146–157.
McCagg, W. O., Jr. (1975). *A Szocialista Gazdaság Fejlődése Magyarországon. 1945–1968* by I. T. Berend; *A Magyar Gazdaság Száz Éve* by I. T. Berend & G. Ránki; *Hungary: A Century of Economic Development* by I. T. Berend & G. Ranki, Eds.). *Slavic Review, 34*(4), 857–858.
Murber, I. (2005). A bécsi Collegium Hungaricum újjászervezése (1963). *Levéltári Szemle, 55*(4), 26–40.
Pach, Z. P. (1955). Tanulmányúton a Szovjetunióban. *Századok, 89*(2), 241–261.
Pach, Z. P. (1961a). A XV–XVIII. századi agrárfejlődés egyes kérdései a stockholmi történészkongresszuson. *Századok, 95*(2–3), 226–235.
Pach, Z. P. (1961b). A magyarországi agrárfejlődés elkanyarodása a nyugat-európaitól (A feudalizmusból a kapitalizmusba való átmenet magyarországi sajátosságainak kérdéséhez). *Agrártörténeti Szemle, 3*(1), 1–9.
Pach, Z. P. (1963). A II. nemzetközi gazdaságtörténeti kongresszus. *Századok, 97*(3), 256–258.
Pach, S. P. (1965). Über einige Probleme der Gutswirtschaft in Ungarn in der ersten Hälfte der XVII. Jahrhunderts. In *Deuxième conférence internationale d'histoire économique/Second International Conference of Economic History Aix-en-Provence 1962* (II. pp. 223–235). Mouton & Co.
Pach, Z. P. (1966). A III. Nemzetközi Gazdaságtörténeti Kongresszusról (München, 1965. augusztus 23–27.). *Századok, 100*(1), 19–26.
Pach, Z. P. (1968). The Shifting of International Trade Routes in the 15–17th Centuries. *Acta Historica Academiae Scientiarum Hungaricae, 14*(3–4), 287–321.
Pach, Z. P. (1969). A IV. Nemzetközi Gazdaságtörténeti Kongresszus. *Századok, 103*(1), 66–81.
Palairet, M. R. (1975). *Economic Development in East-Central Europe in the 19th and 20th Centuries* by B. T. Ivan & R. György. Columbia University Press. *Economic History Review, 28*(3), 554–555.
Pogány, Á. (1999). Business History in Ungarn. In A. Teichova, H. Matis, & A. Resch (Eds.), *Business History. Wissenschaftliche Entwicklungstrends und Studien aus Zentraleuropa* (pp. 77–85). MANZ.
Première Conférence Internationale d'histoire économique/First International Conference of Economic History: Contributions, Communications. Stockholm Aout/August 1960. Paris–Hague 1960.
Ránki, G. (1960). A stockholmi történészkongresszus. *Magyar Tudomány, 67*(5), 757–760.
Ránki, G. (1965). Some Problems of Capital Accumulation and Industrialization in Hungary 1867–1914. In *Deuxième conférence internationale*

d'histoire économique/Second International Conference of Economic History Aix-en-Provence 1962 (II., pp. 701–709). Mouton & Co.

Rév, I. (1979). Mennyire tágíthatók egy diszciplína határai? (Gondolatok az Edinburgh-i Gazdaságtörténeti Világkongresszus kapcsán). *Történelmi Szemle, 22*(1), 181–184.

Romsics, I. (2011). *Clio bűvöletében. Magyar történetírás a 19–20. században – nemzetközi kitekintéssel*. Osiris.

Somlai, K. (2009). Ösztöndíjjal Nyugatra a hatvanas években. Az Országos Ösztöndíj Tanács felállítása. In Tischler, J. (Ed.), *Kádárizmus Mélyfúrások* (pp. 273–314). Évkönyv XVI. 1956-os Intézet.

Somlai, K. (2010). "Kiutazása érdekünkben áll" A nyugatra utazó ösztöndíjasok és a hírszerzés kapcsolatai a Kádár korszakban. In Kozák, Gy. (Ed.), *Kádárizmus Átereszek* (pp. 241–262). Évkönyv XVII. 1956-os Intézet.

Somlai, K. (2021). Amerikából jöttek, hires mesterségük: Fordosok. A Ford Alapítvány ösztöndíjas programjában résztvevők karrieríve: szubelitből az elitbe. In Rainer, M. J. (Ed.), *Elit és szubelit a Kádár-korszakban – közelítések. Magyarország a jelenkorban* (pp. 103–164). Évkönyv XXVI. 1956-os Intézet Alapítvány Kronosz Kiadó.

Takács, R. (2014). Nyitottság és zártság dilemmái – Kulturális érintkezések a külfölddel és a nyugattal. *Századok, 148*(6), 1491–1515.

The Eighth International Economic History Congress (August 16–20, 1982, Budapest) (1983). *Acta Historica Academiae Scientiarum Hungaricae, 29*(2–4), 211–216.

Ujváry, G. (2017). *Kulturális hídfőállások. A külföldi intézetek, tanszékek és lektorátusok szerepe a magyar kulturális külpolitika történetében. II. Bécs és a magyar kulturális külpolitika*. Ráció Kiadó.

Ungváry, K. (2006). A beszervezés és az útibeszámoló. Tudósok az állambiztonság célkeresztjében. *Élet és Irodalom,* máj. 19.

van der Wee, H. (1993) International Economic-Historical Research in Retrospect (1960–1990). In T. Riis (Eds.), *A Special Brew ... Essays in Honour of Kristof Glamann* (pp. 91–105). Odense University Press.

Világtörténet 1968. No, 16. A magyar-francia gazdaság- és társadalomtörténeti konferencia főreferátumai (pp. 3–73).

Zimányi, V. (2001). Une collaboration fructueuse: la mémoîre du professeur Braudel. In P. Sahin-Tóth (Eds.), *Rencontres intellectuelles franco-hongroises – Regards croisées sur l'histoire et la littérature* (pp. 39–44). Collegium Budapest Workshop Series No. 10.

CHAPTER 6

Poland: Economic History Debate Across the Iron Curtain

Anna Sosnowska

Polish post-1945 history departments were devastated by war atrocities, deaths of faculty members, and destruction of archives and libraries. Additionally, some historians who were well-established in prewar historiography survived the war but opposed the Soviet domination in Poland and bitterly decided to stay in exile. Even the most prolific and famous of them, Oskar Halecki in exile in New York City, ended up isolated from both colleagues in his country of origin and mainstream historical debates in his new country's academia. Social sciences including historiography became dominated in post-1945 Poland by Marxism. Especially in the Stalinist early 1950s, the communist state's ideological censorship limited academic freedom. Yet, the late 1950s and 1960s was a golden period of Polish economic historiography unmatched by any other period in terms of both output and international cooperation and influence. Polish scholars established intensive cooperation with the

A. Sosnowska (✉)
University of Warsaw, Warsaw, Poland
e-mail: aksosnow1@uw.edu.pl

A. Dolezalova and C. Albrecht (eds.), *Behind the Iron Curtain*, Palgrave Studies in Economic History,
https://doi.org/10.1007/978-3-031-31578-7_6

French Annales school of global economic history. The research created in this period, especially works by Marian Małowist and Witold Kula, was impacted by and had influence on Western left-wing debates on the origins of capitalism and the East-West economic divide of Europe, including publications of Fernand Braudel, Immanuel Wallerstein, Perry Anderson, Robert Brenner, and Ivan T. Berend.

What circumstances can explain the intensive exchange and relative influence of Polish economic historiography unmatched by other disciplines at the time? This text aims to provide an explanation for this unique development by tracing historians' flow of ideas in their works as well as in their biographies, memoirs, letters, and interviews. My answer takes also into account the extensive historical literature on and sociological interpretations of the development of postwar historiography in Poland.

The postwar breakthrough stemmed from institutional changes introduced in academia by the newly established pro-Soviet regime with Marxism-Leninism as its official ideology. Historical materialism not only set the political goal of heading toward a classless communist society led by the communist party (hereafter the Party) but also demanded a reinterpretation of the past. History was seen as the story of technological change, the struggle between the dominant and oppressed social classes, and the evolution through social economic formations from feudalism to capitalism to socialism. Cold War competition between the capitalist West and the socialist Soviet bloc made justifying the superiority of the latter over the former the sole obligation of historiography. Ideological arguments were supported by the results of historical research. The postwar shift in historical studies in the West was also a factor of change, even if less clearly recognized, in Polish historiography. With nationalism discredited by the Nazi crimes and Marxism introduced as the official ideology, Polish historians focused on the history of large social groups other than nations and dynasties; material culture and economic history rather than the history of great individuals and national heroes; and political institutions rather than wars, treaties, and diplomacy.

As I document below, in the postwar period, economic historians' research continued its focus on the divergent paths of economic development in Western and Eastern Europe since the turn of the Middle Ages and early modern period (Kochanowicz, 2016). Yet, most influential postwar works were novel in their explanatory emphasis on patterns of international trade and the global division of labor as well as contextualizing East-West European relations within the West's overseas expansion and the rise of a capitalist center in northwestern Europe in the modern period (Sosnowska, 2019).

Mostly with Marx But Not Entirely with the Party

From the early postwar period through the 1980s, historians and their institutions in Poland developed cooperation across the Iron Curtain. In postwar Poland, economic and social history gained a stronger position than it had held in the interwar period. It was able to secure this strong position all the way to the 1970s when economic and social history slowly but surely gave way to cultural history and the history of ideas. The strong position of economic historical research, local and international funding, prominent institutions, and representation at international congresses and Western academic journals as well as extensive international contacts was a result of both the internal dynamic of Polish historiography and the global dynamic of historical studies.

Out of six pre-World War II universities, only four remained within the borders that Poland was assigned at the Yalta conference of 1944: the University of Warsaw, the Jagiellonian University in Krakow, Adam Mickiewicz University of Poznań, and the Catholic University of Lublin. Two universities remained outside of Poland's postwar borders. Lviv University, in Ukraine since 1945, where humanities, including philosophy and economic historiography (led by Franciszek Bujak) had had a strong position in the interwar period. Finally, Vilnius University became a part of Lithuania after 1945. Yet, briefly after the war, three new universities were created by the communist government with the idea that they would model new research and faculty members and provide education to the lower classes, which supported the new socialist society: in Lublin, Łódź, and Toruń (Stobiecki, 2020). The Wrocław/Breslau university was also incorporated into the new education system.

The University of Warsaw lost the largest share of its faculty members. Executions, deportations, extermination of Jews and intelligentsia, emigration, war poverty, and undernourishment resulted in 53 percent loss in the number of historians at University of Warsaw (Romek, 2010). In the entire country, 19 percent of all university historians employed before the war died during the war (Rutkowski, 2007, 23). According to a presentation on the state of the historical studies in the First Congress of Polish Science in 1950, 54 history committees in seven universities employed over 150 faculty members, including 52 professors, of whom almost one-half (24) had obtained their positions after 1945 (Kormanowa, 1951).

As a result of the sophisticated game that the professors of history educated before World War II played among themselves and with the representatives of the new Marxist orthodoxy imported from the Soviet Union, economic history as a subdiscipline and economic historians as individual participants of the academic life gained a hegemonic role in the institutions of historical research especially between 1956 and the 1970s (Behr, 2022; Connelly, 2000; Górny, 2013). Detailed studies of postwar Polish historiography provide an interesting picture of the rivalry and alliances among academic historians with and without prewar university employment, the Party officials responsible for the Stalinization of the academic life, and Soviet representatives overseeing the process (Stobiecki, 2020). From 1948 to 1955 a group of historians, disproportionately based in Warsaw, interested in economic history, and of Marxist convictions rooted in their perception of the Great Depression managed to dominate historical studies in Poland through the 1970s. As indicated by the prosopographic analysis of Valentin Behr (2022), the core of this group, including economic historians Marian Małowist and Witold Kula, obtained their advanced degrees and demonstrated research excellence before the war, enjoyed the respect of the prewar left-wing historical milieu, and after the war were employed at the University of Warsaw's departments of history and economics and in the Polish Academy of Sciences (hereafter the Academy) (Behr, 2022, 55). As demonstrated by Maciej Górny (2013), Rafał Stobiecki (2020, 57), and John Connelly (2000), the prewar professional historians' leading position in historical studies set Poland apart from other countries on the Soviet side of the Iron Curtain. In 1950, after five years of Stalinization attempts, only seven out of 52 history professors in Poland were Party members (Behr, 2022, 36). Even among the group of 47 historians that Behr identified as playing a major role in the ideological and institutional Stalinization of Polish historical studies in the early 1950s, only nineteen were Party members. Out of 25 core members of the "Stalinization squad," barely a majority (13) belonged to the Party (Behr, 2022, 38). What is more, "commonly regarded as the symbol of the Stalinization of the Polish history writing, a famous Otwock conference organized (...) at the turn of 1951 and 1952, ended with a defeat of the most dogmatic group of historians-Marxists."

Despite the dramatic changes in Poland's borders, population, and material academic infrastructure during the war, followed by establishment of the Soviet-style political order, economic history research maintained its connection to its prewar roots (Kochanowicz, 2016). Franciszek Bujak (1875–1953), a prewar professor of economic history at Lviv University, continued his research at the Jagiellonian University's Department of Agriculture in Kraków. Jan Rutkowski (1886–1949) still headed the economic history division at the University of Poznań's history department. The two of them were able to reactivate the major journal of economic history *Roczniki Dziejów Społecznych i Gospodarczych* [*Social and Economic History Annals*, hereafter *Roczniki*], which they had established in 1931 in Lviv, inspired by *Annales: économies, sociétés, civilisations* (Guzowski et al., 2015). *Roczniki* inaugurated its postwar life in 1946 with an issue interestingly labeled 1939/1946, which collected articles reporting prewar, wartime, and fresh postwar research. It has remained the most important journal of economic history in Poland, which bridged the prewar and new research centers as well as subsequent generations of economic historians up to the present. Poznań also continued in the postwar period as a crucial hub of research in economic history and a place where *Roczniki* was published until 1996. Edited initially by its founders Bujak and Rutkowski, it was taken over by the latter's students Władysław Rusiński, whose tenure lasted for three decades until 1987, and Jerzy Topolski. In 1967 economic historians from the Poznań school, Jerzy Topolski, and Czesław Łuczak, founded another specialist journal, *Studia Historiae Oeconomicae*. Published in English, it aimed "to share research results of Polish scholars in the field of economic and social history to international readers."[1]

Yet, it was the University of Warsaw—and none of the newly created universities dominated by Marxist historians—that became the main center of historical economic research and international cooperation, focused at its peak around Marian Małowist and Witold Kula. Kula started his postwar career at the newly created University of Łódź. His research developed at the University of Warsaw's Department of Economics, where he headed the economic history division. The fact that both of them had conducted substantial research before the war, combined with

[1] https://pressto.amu.edu.pl/index.php/sho/about.

the strength of their leadership all the way to the 1980s, added to the continuity of economic historical research. Therefore, in the long run of the Cold War, economic history appeared to be peacefully succeeding Bujak's and Rutkowski's traditions rather than building a postwar Marxist scholarship in opposition to it. Although the exact number of researchers that could be qualified or thought of themselves as economic historians is not known, it is possible to reconstruct the geography of the subdiscipline via analysis of *Roczniki*'s editorial board throughout the Cold War. This analysis confirms the primacy of the Warsaw and Poznań schools of economic history, the stable if modest position of Kraków and Wrocław, and the rise of Gdańsk at the end of the Cold War era.

The respect for historical research traditions combined with new leadership became characteristic not only of the economic history milieu but for postwar historical studies as such. The difference between the two could be described as a difference in degree rather than in content. The process of installing leadership loyal to the Party required more brutal interventions in the field of general history. New professorship nominations for Party members were not the only way the new communist government planned to reconstruct the universities and especially the social sciences. The prewar system of universities and research institutions became centralized and controlled by the state. Non-governmental academic associations were barred or marginalized during the early postwar years (Stobiecki, 2020, 53–54). Universities developed new programs of history studies and introduced ideological courses of Marxist-Leninist political economy and historical materialism.

In 1953 a major research institution free from teaching and directly connected with the government, the Polish Academy of Sciences (hereafter, the Academy), was established. It was to gather leading researchers in each field even if they were employed at universities and set the main directions of the research. Its Institute of History played a crucial role in organizing historical research and publications through the entire Cold War period (Stobiecki, 2020). The Institute of History never had a separate division of economic history but did establish a social history one. Its structure referred to the politics-based periodization characteristic of prewar Polish historiography. Since its creation, the Institute of History has been responsible for publishing *Kwartalnik Historyczny* [*Historical Quarterly*], the oldest Polish scholarly journal (established in 1887 in Lviv) and one of the most influential academic journals of general history.

Additionally, the Academy included also an Institute of History of Material Culture, a sub-discipline never recognized before the war. With roots dating back to the post-Yalta acquisition of the formerly German land and a research committee devoted to investigating Slavic traces there as well as preserving the abandoned German heritage, the institute employed archeologists, ethnologists, and ethnographers along with historians. (Significantly, the institute was renamed the Institute of Archeology and Ethnology in 1992). Since 1953 it has published *Kwartalnik Historii Kultury Materialnej* [*The Quarterly of the History of Material Culture*], which included articles authored by economic historians Wyczański and Kula, to mention only a few.

The Institute of History dominated not only historical research but also international exchange once the latter developed after the political liberalization of 1956. The period of intensive exchange with Western research institution overlapped with the "golden period" of the Warsaw-based school of economic history that lasted from the late 1950s to the 1970s. The year 1956 marks liberalization in the political history of Poland. The era began with workers' strikes in Poznań and de-Stalinization within the Party leadership. After this breakthrough, cooperation with their colleagues in France, and to lesser extend with Great Britain, Germany, and the United States through the Ford Foundation, developed (Kilias, 2021; Kula, 2010; Pleskot, 2010). The existing studies document the primary position of France and the Annales school as a destination of historians' academic trips, intellectual fascinations, and research inspirations (Pomian, 1978; Wierzbicki, 2004). Pleskot claims that "in 1965 about 188 representatives of the humanities traveled to capitalist countries via the 1st Department [including history] of the Polish Academy of Sciences (...), including as many as 68 to France, 20 to Austria and Italy each, 10 to Scandinavian countries and Great Britain and 7 to the U.S." (Pleskot, 2010, 33). Marcin Kula (born in 1946), a historian and a son of historian Witold Kula and sociologist Nina Assorodobraj-Kula, asserts in his autobiographical book *Mimo wszystko bliżej Paryża niż Moskwy [In spite of everything, closer to Paris than Moscow*] that a significant part of cooperation between Polish and French historians was tied up in "[my] parents' home. Especially [my] father was a crucial wheel in the mechanism of this relationship" (Kula, 2010, 9). The Warsaw-based historians whom Behr identified as dominating historical studies since the early 1950s (economic historians Witold

Kula, Marian Małowist and his students, as well as medievalists Aleksander Gieysztor and Tadeusz Manteuffel) and their students animated the exchange and benefited the most from the cooperation (Kula, 2010; Pleskot, 2010, 158–162, 170–172).

According to Pleskot's estimations of the number of months that Polish humanists and historians spent at the Annales school, the period between 1959 and 1967 brought the highest turnout. Between 120 months (in 1959) and 189 months (in 1965) were granted during 60 to 100 visits. The peak years of academic travels were 1963 to 1965. Historians represented the largest group of scholars and made up one-fourth to one-third of all Polish visitors (Pleskot, 2010, 33). Although French historians organized the network of Western and Eastern economic historians, American foundations financed Polish scholars to a higher degree than did the French government (Berg, 2015). Rockefeller Foundation grants and American cliometric economic historians helped sustain the International Association of Economic History (hereafter the IEHA) (Berg, 2015). The pattern of Western assistance and influence shaped historical studies differently than sociology. The latter remained under the American spell both intellectually and financially. The activities of the Ford Foundation intensified after the 1956 liberalization but attracted Polish sociologists to U.S. research centers and scholarship. "The Ford Foundation program, which started in 1957, consisted of four grants used mostly to finance the academic travels of Polish scholars who wished to visit Western academic institutions. It dealt not only with the United States, but also with Great Britain, France, Germany, and a few other countries. (…) [Yet] overall, the Polish-American exchange was the largest and the most important one" (Kilias, 2021, 34).

Polish economic historians also took part in meetings of the International Committee of Historical Sciences. Although absent from the first postwar meeting in Paris in 1950, the Polish representation prominently included economic historians in the congresses organized in Rome in 1955 and later. International historical congresses provided also rare occasions for historians from Poland and exile Polish historians to meet and for both to realize the ascending importance of economic history in postwar historical studies (Zadencka, 2015).

Polish economic historians were able also to add to the academic debates via publications in congress languages in *Annales*, the *Economic History Review*, the *Journal of European Economic History*, and *Past and Present*. Kula's *An Economic Theory of Feudal System* became a champion of translation among books published and discussed in the 1970s and 1980s in Italian, Spanish, French, English, and Portuguese.

Open Window: 1956–1980

The three subsequent subsections discuss dominant themes in Polish research and the thematic connection with research in other countries. *The Annales's Polish connection* stresses the role of a window to the international intellectual world that the Parisian center played for Polish scholars during Braudel's presidency, especially in the 1960s and 1970s. The section *Grain trade, serfdom, and capitalism debate* analyzes the mutual if unequal impact of Polish and international scholarship on the analysis of the "second serfdom" in Eastern Europe and its influence on international trade in East-West European economic relations, capitalism, and colonial expansion in the early modern period. Finally, *Historians in hostile isolation* provides a closer look at the complicated biography and academic legacy of Oskar Halecki, a war refuge, anti-communist historian based in New York City, and alleged author of the term "East Central Europe." Juxtaposing his story and work with Małowist's, his former student based in Warsaw, I demonstrate how their rival depictions of East (Central) European development at the turn of the modern era were rooted in their divergent life experiences and ideological standing.

The Annales' Polish Connection

The Poles, like other East Europeans but in higher numbers and intensity, had strong connections with the Annales school, especially under the leadership of Fernand Braudel (Kamecka, 2004; Górny, 2021). They also participated in the international congresses organized by the IEHA (Bębnowski, 2019; Berg, 2015, 37). Recent Polish literature suggests that no significant intellectual results followed the Polish trips to the Annales school. Pleskot's (2010) and Kula's (2010) large volumes, which are devoted entirely to the exchange between Polish historians and the Annales in the postwar era, argue that the two were intellectual neighbors that under the influence of the same *zeitgeist* provided Polish researchers who otherwise would have been isolated with "a window to the world" (Kula, 2010). Yet, they conclude that "the methodological influence of the Annales school was limited" (Kochanowicz, 2016, 219) and vague (Górny, 2021; Pleskot, 2010). Cooperation allegedly took the form of research visits to academic libraries and personal friendships (Mączak & Tygielski, 2000, 167).

My conclusion is different. Intrigued by these opinions, I combined my old research with the new one to investigate the flow of the ideas. I conclude that—contrary to the popular opinion formulated by Pleskot and reproduced in influential sources—there was a meaningful intellectual exchange, apart from general academic benefits. Its dimensions have been acknowledged and pointed out by the historians who participated in the cooperation, including Andrzej Wyczański, a leading postwar economic historian (Pomian, 1978; Guzowski et al., 2015). Additionally, I traced the flow of ideas in the works of authors such as Małowist and Kula. These three historians were the main actors of the exchange on the Polish side, while Braudel and Immanuel Wallerstein were the main actors on the Western side. The stages where the action of the exchange took place were institutions coordinated by Braudel: the Annales school in Paris, IEHA congresses, and economic history seminars at the Datini Institute of Pre-Industrial Economic History in Toskanian Prato.

Scholarships and congress participation both were dominated by Małowist, Kula, and their students. In the 1970s Andrzej Wyczański and Jerzy Topolski, their opponents in interpretations of East-West trade relations in the early modern period, joined the international debate (Topolski, 1974a, 1974b; Wyczański et al., 1985). The intellectually competitive and personally distant relationships that these four researchers developed are famous even if discursively moderate in their memoirs, letters, and biographies (Kula, 2010, Siewierski, 2016). While commenting on a photograph depicting the two power couples—his parents and the Małowists on the Eiffel Tower in 1947—Kula's son ironized that "in some time ahead, such a configuration would unfortunately have been difficult to replicate in a context of friendship" (Kula, 2010, 388). Kula characterologically and politically disapproved of Topolski (Kula, 2010, 73, 429, 524, 672), who served as a board member of IEHA in the 1970s after the partial retirement of Kula.

The four historians cooperated with the Annales in their own ways despite the common traits they shared. Three out of four main players were economic historians based in Warsaw. Born within a span of 20 years, they shared an interest in the Polish economy of the early modern period. While Wyczański and Topolski conducted some research and publication as a team, each of them benefited from their contacts with the Annales differently.

The oldest of the four, Marian Małowist (1909–1988), was born in Łódź into the family of an assimilated Jewish physician. Although an ambitious and appreciated researcher, in the interwar period he had no chance for a university position and worked as a teacher in a Jewish high school. Up until the outbreak of the Second World War, Małowist collected materials for a monograph about the Genoese colonies on the Black Sea in the fifteenth century. During the war, having escaped from the Warsaw Ghetto, he remained in constant hiding, teaching at a village school in the Podlasie region. Throughout the war years he also continued his research on Italian trading colonies, which led to his postdoctoral thesis on Kaffa, a fifteenth-century Genoese colony on the Black Sea, which he defended right after the war.

Marxist by conviction and close to the communist youth organization as a teenager in Łodź, Małowist never signed up for Party membership. The postwar period brought a rapid improvement in his academic status. Employed by the University of Warsaw in 1946, by 1955, he had become a full professor and head of the Department of Medieval History (Siewierski, 2016). He played a key role in the postwar era, including the hard-core Stalinist reorganization of historical research and teaching, always as a solid economic history researcher and methodological Marxist and in tactical cooperation with Party activists and Soviet advisors (Siewierski, 2016, Stobiecki, 2020).

Małowist was interested in the great global changes associated with the expansion of Europe, the beginnings of capitalism in the West, and the re-feudalization of Eastern Europe in the late Middle Ages. He argued that the early modern era determined the future development paths of Eastern and Western Europe, which were separate yet complementary; it also determined the modern system of economic exchange and international division of labor. In his never-explicated and yet clearly presented model, early modern developments prepared the way for northwestern Europe's global economic advantage in the nineteenth century. Once, it had achieved economic advantage over regions with which it maintained relations, Europe developed by expanding into those very same regions, the first of which was Eastern Europe.

Particularly when he was producing wide-ranging studies in the 1970s (Małowist, 1969, 1974; Małowist et al., 2010, Bieżuńska-Małowist & Małowist, 1987), Małowist's approach to social change was global in character. In his view, there was an obvious link between the working conditions of a Polish weaver, on the one hand, and the complex interrelationships among the structure of the grain trade in the Baltic and North Sea regions, the new trading opportunities for West European merchants following the conquest of West Africa, and the expansion of Europe and the evolution of the quasi-colonial character of the East European economy on the other. This was a similar style of thinking to that of Braudel, Eric Hobsbawm, and Wallerstein. All three of them played a role in Małowist's methodology, interpretation, or the reception of his works. Although Małowist visited the Annales school in the 1950s and corresponded with co-founder of the IEHA Postan and Hobsbawm, it was his students Henryk Samsonowicz, Antoni Mączak, and Bronisław Geremek along Kula and Wyczański who became "well embedded by 1965" (Berg, 2015, 57) in the delegations to the IEHA congresses, Braudel's seminars, and Prato meetings. Małowist, who remains in historians' memory as a charismatic teacher but not a charmer in peer relationships, received and performed influence on the international debate via written and published works rather than personal contacts.

Yet Małowist, who was working on the same thirty-year period of history as the dependency theorists, came to similar conclusions. Although he probably did not read their work, and his interests were strictly historical and academic, he drew on the same Marxist theoretical and ideological sources. Małowist dealt with other regions, but his model of the relations of Eastern Europe and West Africa with Western Europe is strikingly similar to that of the dependency theorists. They held similar views about the origins of capitalism, the mechanisms by which dependence was reproduced, and the narrowly specialized nature of economic development as a cause of stagnation. Despite the similarity of their arguments, however, it is not possible to uncover a precise link between the Dependencia school and Małowist's research (Małowist et al., 2000). Kula, on the other hand, made direct use of the Dependencia theorists' research, to which he was introduced by the younger generation of Polish researchers mentioned above. His categorization of Eastern Europe and Latin America as "raw material supply bases" for the West was an idea he developed from the mid-1960s.

Małowist's junior by a few years, Witold Kula (1916–1988) received education in both history and economics. Unlike Małowist, he had a wide-ranging interest in the social sciences. Just before the outbreak of the Second World War he defended his PhD dissertation on the demography of the nineteenth-century Russian empire's Kingdom of Poland, while in 1947 he completed his habilitation thesis *Social Privilege and Economic Progress*. He worked in several academic institutions in Łódź and Warsaw before landing for life at the economics department of the University of Warsaw and the Institute of History of the Academy. He chaired economic history in the former, while in the latter he co-led the research team focused on nineteenth-century social structure.

From 1956 Kula was an active participant and moderator of the research visits to the Annales school. He was an unquestioned leader of the Polish contingent there (Berg, 2015; Kula, 2010; Pleskot, 2010). That is confirmed by Polish research (Pleskot, 2010), Marcin Kula's memories of his parents' home office, and correspondence networks. Kula was the informal decision and distribution center of the scholarships provided by the American-French economic history establishment (Kula, 2010). His influence is also reflected in the formal positions that Kula held within this establishment. He visited the Annales more frequently than others, regularly participated in the IEHA congresses, and became president of the IEHA in 1968–70 and an honorary one after his health deteriorated in the 1970s. Finally, he held his own seminar at the Annales school in 1963–64 and 1971–72. Although he also lectured in Italy, Switzerland, and Czechoslovakia (Stobiecki, 2020, 23), it was the Parisian Annales school that provided him with a wide-open window to the world.

Because of the frequency and length of Kula's trips, the private relationships and friendships that Antoni Mączak (Mączak & Tygielski, 2000) overemphasized as the main result of Polish-French cooperation were stronger for him than for others. He often traveled with his wife Nina Assorodobraj-Kula (Kula, 2010), whose research also benefited from American scholarships in France. Therefore, his social position was stronger than other East Europeans. In the overwhelmingly male world of economic historians of the 1950s and 1960s, Berg observed a difference between Western and Eastern scholars. While the former were able to consolidate their academic relationships from Prato seminars during the "dances outside the formal events and at frequent meetings between the congresses" (Berg, 2015, 68), the latter were typically deprived of this opportunity as their wives did not accompany them because of visa restrictions and cost. What made the Kulas a power couple in the Polish social science milieu (Behr, 2022) also favored Kula's social position in international economic historians' gatherings.

Kula was also more successful than Małowist in disseminating his works in the international publishing market. Unlike Małowist, whose research results appeared in journal articles in both Polish and foreign languages, Kula's ideas found better expression in book format (Kula, 1956, 1976, 2001). His record of books published in foreign languages in the 1960s and 1970s is not only impressive even by today standards but also unprecedented and exceptional in the conditions of Cold War era. Kula's network tied in Braudel-patronized events and institutions as well as the individual charisma and influence of the French "pope of historiography." Braudel's decision to share his influence with Kula in the introduction to the French edition of *An Economic Theory of Feudal System* (1970) seems to have played a crucial role in further publications of the book. Subsequently, the book appeared twice in Spanish, including as soon as 1972 in the Argentinian edition, then in Italian and in English in the renown leftist Verso in 1976, in Portuguese in 1979, and in Hungarian in 1985.

Kula's relationship with Braudel in terms of personal relations and mutual intellectual influence unfolded differently than Małowist's. Although Kula often cited Braudel (Pomian, 1978), other international influences are also manifest in Kula's writings: the methods of American cliometrics, economic modeling, the postwar Dependencia and modernization theory that inspired American debates on the development in the Third World, and French Marxist anthropology (Sosnowska, 2019). Yet, there is enough evidence (Berg, 2015; Kula, 2010; Pleskot, 2010; Wierzbicki, 2004) to claim that these international influences were made possible by social contacts established at the Braudel-dominated Annales school, IEHA congresses, and Instituto Datini in Prato in the 1950s and 1960s and at the Parisian libraries that Kula had a chance to explore during his numerous stays at the Annales. As Berg demonstrates, in Paris Kula met Alexander Gershenkron, David Rostow, David Landes, and Michael Postan, all of whom fascinated him and influenced his works (Kula, 2010, 616). Marcin Kula's collection of family letters and recollections testifies to the significant amount of time that Kula senior spent in Braudel-patroned institutions in France and Italy as well as at French and Italian libraries, archives, and museums. Kula's student Tadeusz Lepkowski's observation that 75 percent of the examples in Kula's works come from Polish, French, or Italian history (Łepkowski, 1990), and my own discovery of Kula's geographic imprecision (Sosnowska, 2019) comes as no surprise given the degree to which Kula's research experience was connected with these three countries.

Although Kula's and Braudel's relationship was asymmetrical when it came to the influence that it had on their respective careers, associated funds, and prestige, it was far from one-sided. Kula was instrumental in the popularization of Braudel's works among a wider audience than university seminars held by Małowist and Kula and citations by Braudel's visitors. Soon after the French edition of Kula's *An Economic Theory*, a Polish translation of Braudel's *Historia i trwanie* [*On History*] was published in 1971 and *The Mediterranean* in 1976, both with introductions by Kula and Bronislaw Geremek. Geremek was a historian of the medieval Paris underclass, originally a student of Małowist, a friend of Duby and LeGoff, and a key Polish politician in the 1990s.

Andrzej Wyczański (1924–2008), another Warsaw-based historian, carved a relatively independent mode of cooperation with the Annales. "Not a celebrity type of a researcher" represented by the generation-older Kula, the Annales school's influence on him was less spectacular but systematic and self-analyzed (1999). Wyczański, while a specialist in the manorial and peasant economy and food consumption, also investigated the history of diplomacy, public administration, domestic policy, social stratification, the education and history of the Jagiellonian University, and symbolic culture, including beliefs and religion in the Polish–Lithuanian Commonwealth (Wyczański, 1960, 1964, 1965, 1970, 1987). Wyczański was a member of the first group of younger scholars who benefited from the post-Stalinist exchange between Poland and France and visited the Annales school first in 1956. He returned repeatedly in the 1960s and held his own seminars in the 1970s (Wyczański, 1993, 1999). Focused on the sixteenth-century history of Poland, this self-described follower of the Annales paradigm of integral history (Guzowski et al., 2015, 32, 49) looked at his subject from numerous perspectives. Although his publications avoided methodological notes, he described himself in a 1990s interview as a scholar whose ambition was to reconstruct an era's mentality by studying institutions, documents, and other artifacts. His most-cited book, *Polska Rzeczą Pospolitą Szlachecką [Poland—A Res Publica of the Nobility]*, "was very much rooted in the concepts of Ecole des Annales; I mean here especially approach to culture (...) [understood as] mentality and not objects" under "the influence of Lucien Febvre through Robert Mandrou" (Guzowski et al., 2015, 85).

Finally, Jerzy Topolski (1928–1998) represented the Poznań school of economic history in international meetings of economic historians. Unlike the three others, he was a Party member from 1948 and he published in foreign languages—Italian and English rather than French like Kula. Topolski experienced the Anglo-Saxon turn in economic historiography and was influenced by British Marxist historians. Although he actively participated in the "grain, serfdom, capitalism" debate, ultimately, his Polish and international legacy was in historical methodology rather than economic history (Norkus, 2012). Jerzy Topolski's most famous book was *Methodology of History*. Published originally in 1966, it appeared in Italian in 1975 and in English in 1977. Receptive to the debates in European and American historiography, Topolski experimented between the 1950s and the 1980s with combining various versions of Marxism with the Polish prewar research traditions. His interest in the latter traced back to the scholarship of Jan Rutkowski, who mentored Topolski in the 1950s. Topolski picked up Rutkowski's focus on early modern East European agrarian economy and his inspiration from and contacts with the Annales. Although a fan of modeling, like Kula and Wyczański, Topolski's inspirations cannot be so clearly associated with the Annales. This later version of his model is a response to the challenges presented by Western as well as national historiography (Topolski, 1982). In the 1980s, Topolski engaged in polemics with Anglo-Saxon researchers Maurice Dobb, Perry Anderson, Richard Brenner, and Immanuel Wallerstein first and foremost (Topolski, 1982, 1983). Topolski served as a board member of IEHA in the 1970s, participated in its congresses and visited frequently, and held series of his own lectures at the Annales in 1977 and 1981 (Stobiecki, 2014, 348).

In Polish history studies in the 1960s and 1970s, the official version of historical materialism became marginalized. Among intellectuals, including Kula, faith in socialist ways of overcoming backwardness also decreased (Kula, 1996, 415–417). The perspective and dilemmas of dependency theory thus became, especially in Poland and Hungary, an attractive intellectual proposition—more so in economics and historiography than in sociology. Kula collaborated with Iván T. Berend and György Ránki, whom he met at IEHA meetings (Berg, 2015) and who from the late 1970s were influenced by dependency theory and Wallerstein's world-systems approach. Despite the similarities in their understanding of the region's history and their similar ideological inspiration, Małowist was a less-frequent participant in French and Italian meetings organized by Braudel and did not establish contacts with the Budapest milieu (Siewierski, 2016). The Polish historians of Latin America Marcin Kula, Tadeusz Łepkowski (both close to Witold Kula via family and mentorship ties), Ryszard Stemplowski, and Henryk Szlajfer also adopted a position similar to that of the Dependencia school on the development opportunities of undeveloped countries (Sosnowska, 2019, 179). The Annales research stays and stipends were crucial for the rise of this direction of studies. Marcin Kula recollects meeting Celesto Furtado, a Brazilian economist in exile, and relying on Braudel's network in raising funds for his research trip to Brazil. M. Kula especially appreciated the access to literature and primary documents in Paris libraries for his studies of the Cuban revolution and African slavery in Brazil (Kula, 2010, 339). Łepkowski stayed in Paris with Annales' research scholarships through the late 1950s. Paris, in the 1960s the capital city of a falling colonial empire in Africa and a haven for political refugees from Latin America, was instrumental in fostering the interests of Polish social scientists in developing countries and the Dependencia school. Braudel's networks and funds also facilitated Polish scholars' trips to Latin American countries.

Grain Trade, Serfdom, and Capitalism Debate Across the Iron Curtain

The most influential Polish economic history studies focused on explaining Polish and East European relative (vis-à-vis the West) backwardness and divergent developmental paths in the East and West of Europe. Yet, Polish economic historians took part in, were impacted by, and influenced the debate that organized much of the economic historical research since World War II.

While Polish historians' research obsessively investigated the question of East European economic weakness vis-à-vis the West, Western scholarship saw it as a debate on the historical origins and nature of capitalism, which in its view had developed nowhere else but the West and provided it with the opportunity to rise above other parts of the globe.

In many ways, the debate anticipated more recent debates in global (economic) history, captured again by historians in the United States—on the Great Divergence and especially the New History of Capitalism with its focus on the role of slavery in the development of global capitalism. The 1960s and 1970s debate bears some similarities to the current debate. It was, however, framed differently. Polish historians observed the global capitalist economy from an East European corner of the world. Unlike today, they made the region's past crucial to the debate on global capitalism. The earlier debated focused on Europe, the attachment to grain rather than cotton, and serfdom along with the New World slavery.

Małowist studied, reviewed (1967), cited, and recommended to his students Braudel's works. The assignment of *The Mediterranean* is often quoted by former students terrified at the time by the volume of reading and the deadline for the assignment (Siewierski, 2016, 10). Braudel's influence is most clearly visible in Małowist's respect for and attachment to geography. Given Małowist's fascination with Braudel's *Mediterranean* and the scarcity of Małowist's personal recollections (referring, e.g., to motivation and inspiration), I have a good reason to think that Małowist's regional approach to historical studies and his focus on international trade routes (visible already in the 1930s) consolidated in the 1950s, after Małowist's thorough studies of *The Mediterranean* (1949).

Małowist's interest in the Baltic sphere, Balkan sphere, central European sphere, Scandinavia, and west Africa make a collection of peripheral regions of the early modern European economy that were linked by international trade routes. Long-distance trade organized by European merchants among the Old World, the New World, and Asia, as in Braudel's analysis, united various regions and their divergent assignments making them complementary within the economic system. Given his early prewar publications, I am certain that Małowist's idea to follow the routes of international trade precedes *The Mediterranean*. Yet, the way his magnum opus *Wschod a zachod Europy* [*East Versus West of Europe*] (2006) is organized, with chapters focused on regions specializing in particular raw material provision to the European-led economy, and resembles the contents of *The Mediterranean*'s chapters from volume 1.

Małowist's fascination with international trade and his geographically broad perspective on the world economy had much in common with Braudel's vision, which was fully expounded in *Civilization and Capitalism* (1992). In the 1960s, Małowist had already attributed the origins of capitalism to the international trade links subsequently identified by Braudel as the outcome of Europe's global expansion from the fifteenth century onwards. Yet in a review of the first volume of *Civilization and Capitalism*, Małowist challenged Braudel's idea that the metropolis had played a leading role, revolutionizing economic life. He pointed out that in the Middle Ages it was crafts, usually practiced in the countryside and in small towns, that had played this role (Małowist, 1967). The most striking feature of Małowist's vision was that he categorized economic entities in terms of the geographically defined economic area to which they belonged. This categorization was implicit and was not accompanied by a methodological commentary, but it was clear. Like Braudel, for Małowist an economic area had to have some sort of objective creative force that was the outcome of its natural qualities—geography, climate, and demography. This objective potential was then transformed (maximized or minimized) by history. Małowist's work along with Braudel's had crucial impacts on the seminal world-systems theory developed by Immanuel Wallerstein in the 1970s (Wallerstein, 2011). Although written by a sociologist and based on the historical work of others, Wallerstein's three-volume history of modern world-systems criticized capitalism for exploitation of the peripheries by the core and for the ever-increasing inequality between regions and peoples and presented his argument with "explicit political aspirations." The first volume of the series *The Modern World-System* is dedicated to Braudel and Małowist. While Wallerstein visited the Annales school, corresponded with Braudel, and named after him the Center for the Study of Economies, Historical Systems, and Civilizations in 1976 (closed in 2020 after Wallerstein's death) in Binghamton, NY, he never met or established personal contact with Małowist. Wallerstein knew the research of Małowist from the handful of articles the latter published in *Annales, Past and Present,* and *Economic History Review* in the 1950s and 1960s (Małowist, 1958, 1959, 1966, 1974).

In Wallerstein's interpretation, the core of capitalism shifted in the sixteenth century from Italian city-states to Amsterdam, and then in the eighteenth century to imperial Britain, in the twentieth century to the U.S., and later to the Pacific core. The capitalism that arose in northwestern Europe in the sixteenth century developed through the colonial or quasi-colonial (economic) exploitation of successive territories incorporated as peripheries providing raw materials such as labor, food, or mined goods. Eastern Europe and the American colonies were the first such peripheries taken over as early as the sixteenth century.

Wallerstein borrowed from other scholars the historical knowledge he used to produce his work covering four centuries of the history of Europe and its expansion. The approach was explicitly inspired by the Latin American dependency theory. He drew on the publications and style of thinking of the Annales school and especially Braudel's synthesis, which traced globalization along the routes of international trade. Finally, historical studies whose results were published in English- and French-language articles informed his knowledge of regions in subsequent epochs. This was the place where Małowist's studies on the Baltic zone trade and East European grain provisions fell into the scheme of the world-system emerging in the sixteenth century. Wallerstein's interpretation of Eastern Europe as developing differently than northwestern Europe and economically resembling subsequent European colonies in the New World rather than its core societies came entirely from Małowist's research and interpretation of the role the grain trade in the Baltic zone, and then found its way back into the works of Witold Kula and Fernand Braudel.

In Małowist's interpretation, the high Western demand for the products from the East in the sixteenth century prompted a "feudal backlash" in Eastern Europe. The economic system of producing grain was based on the gentry's monopolization of land ownership, the serfdom of peasants, the decline of local merchants and towns, and the region's dependency on the demand for grain from the West. The large landed estates evolved into grain-producing agricultural enterprises, capitalist in nature because they produced for the market. The dominant economic mechanism was as follows: grain produced by serfs for the benefit of landowners and at the request of foreign merchants was transported by raft along the Vistula river to Gdańsk (Danzig), and thence through the Baltic Sea to the cities at the core of the world economy. The trading network within the country was taken over by Baltic merchants. They concluded contracts with the owners of folwarks, extended credit, organized the transport of grain, and distributed imported goods: spices, silk, salt, and wine. Following Małowist (1960), Wallerstein described a trade technique worked out in the sixteenth century by merchants—primarily from Amsterdam—and developed over the next century: "What was the method? Very simple: it involved the purchase of goods in advance of their production, that is payments in advance for supplies to be delivered in the future. This prevented sale on an open market. It allowed the merchants rather than the producers to decide the optimum moment for world resale. And since the money lent tended to be expended by the time of delivery of the goods, if not overspent, the producer was always tempted to perpetuate the arrangement" (Wallerstein, 2011, 122).

Kula's most internationally recognized book, *An Economic Theory of the Feudal System: Toward a Model of the Polish Economy, 1500–1800*, was originally published in 1962, half a decade after the Polish historians' trips to France took off in 1956. It cites Braudel extensively. *A Theory* differs from Kula's earlier books not only in style but also in the historical period covered. Both changed in the direction of the research conducted at the Annales and by Braudel himself. While Kula's earlier (but also later for that matter) monographs dealt with the beginning of industrial capitalism in nineteenth-century Poland, *An Economic Theory* presents a model of the Polish feudal and grain-based economy using calculations—not very precise as critics pointed out (Topolski, 1974a, 1974b)—of output and prices of grain sold in several places by lord and peasant sellers. Kula's cooperation with the Parisian Annales was instrumental in publishing the book in Italian and French with Braudel's forward. The long-term fruit of this cooperation and exchange of ideas turned out to be Braudel's incorporation into *Civilization and Capitalism* of Kula's ideas (mostly from *An Economic History*) about the manorial economy, capitalism in backward countries, and the calorie content of vodka (Braudel, 1992).

There are traces that suggest that the idea of modeling entered Polish historiography via Braudel and Kula. Andrzej Wyczański recollected that he first heard about this method from Braudel around 1957 in Krakow instructing another historian—perhaps Kula—"what the modelling method is about and how it can be used in historical research" (Wyczański, 1999, 163). Wyczanski's 1960 study on the grain manorial estate in the sixteenth century (in which modeling was applied) preceded Kula's *An Economic Theory*, but the handbooks of methodology of history by Assorodobraj-Kula (1963) and Topolski (1976) promoted the approach.

Kula was a specialist in the methodology of history, and he introduced Polish historians to both the theoretical approach to historical research and the concept of the model (Mączak & Tygielski, 2000, 167). His approach to creating research hypotheses, testing them, and determining the conditions under which his model of the Polish feudal economy could be verified was more self-conscious (as shown by his use of figures, for instance) than Małowist's. At the same time, his conclusions had to be more specific and his interpretations of lesser historical and geographical scope. As Kula himself explained, "the more elements we introduce into the model, the richer will be the theory we construct; but the number of societies to which that theory could be applied will decline commensurately" (Kula, 1976, 20). His principal model, presented for the first time in 1962 in *An Economic Theory*, was meant to apply to the Polish economy in the sixteenth-eighteenth centuries. However, Kula also found it to be relevant to other economies, primarily in southern Europe and Latin America. The popularity of *An Economic Theory* especially in

Italy, Spain, and Latin America shows that Kula was able to propose a model that related not only to the Polish economy of the sixteenth-eighteenth centuries. The work presented more general evidence for the way in which economies based on agricultural production and with feudal characteristics (a free or unpaid workforce, ostentatious consumption by landowners) operated during a period when capitalism already existed in other areas and impacted them by creating sales markets, organizing trade, and supplying technological solutions. The terms "mercantile feudalism" and "landed capitalism" were attempts to describe the specific character of backward countries in terms of their deviation from the models developed for countries that were the first to industrialize. *An Economic Theory* gave rise to the concept of "capitalism in a backward country," inspired by Gerschenkron's observations on unequal international competition between early and late industrializing countries also published in 1962. In his dialogue with Wallerstein (which means: indirectly with Małowist) and Kula, Braudel tried to settle the status of the manorial economy not just in relation to the center of the "world-economy" but also in relation to the three levels of economic activity. In answer to the question whether the folwark was a capitalist enterprise, Braudel agreed with Kula that it was not. Folwarks and their owners were market players, participants in mid-level economic activity. The economic calculations of peasants were rooted in the "structures of everyday life" and rarely encroached on the market; if they did, they did so illegally. "Witold Kula's admirable study, which analyzes step by step what the 'economic calculations' of the serf -peasants and their masters might have been in Poland between the sixteenth and the eighteenth centuries," wrote Braudel of *An Economic Theory*, "spells out why the landlords were not 'real' capitalists and could not be before the nineteenth century" (Braudel, 1992/2, 267). Folwarks were nevertheless part of the circuit created and controlled by players at the highest level, which in the sixteenth century meant Amsterdam merchants (Braudel, 1992/2, 93). The chapters entitled "Capitalism and the 'Second Serfdom'" and "Capitalism and the American Plantations" provided "a few significant examples" of "how far capitalism was able to penetrate systems structurally very foreign to it, either by head-on assault, or by dominating production from a distance merely by controlling it at the bottleneck of distribution" (Braudel, 1992/2, 265). Like Wallerstein, or simply because he agreed with Wallerstein's comparison and interpretation, Braudel treated Eastern Europe and areas of Hispanic America in the sixteenth-nineteenth centuries as examples of agricultural economies where the rules of capitalism operated at a certain level. This view also prompted Kula to change his mind about whether the folwark was a capitalist enterprise or not.

At the IEHA congress held in Copenhagen in 1974, two other, slightly younger Polish historians, Andrzej Wyczański and Jerzy Topolski, challenged the interpretation of the sixteenth- to eighteenth-century Polish economy as a quasi-colony dependent on the export of grain. What lasted through the 1960s as a vivid Polish debate on the manorial economy, serfdom, and the role of the grain trade for Polish development and the rise of the global capitalism was exposed internationally only in the year of the publication of Wallerstein's *The Modern World-System I* (2011). Wyczański's and Topolski's picture of relatively autonomous peasants and the autarchic economy of pre-industrial Eastern Europe remained mostly unacknowledged by international economic history until the twenty-first century (Cerman, 2012).

Wyczański and Topolski rejected the colonial thesis and suggested rival models of the rise and demise of the East European manorial economy run by serf labor. Based on the calculations of prices, consumption, and output, which Wyczanski had learned at the Annales seminar by Jean Meuvret (Wyczański, 1999, 162–163), they presented a model of "peasant economy before and at the eve of the industrial revolution," which Wyczański happily recollects as an object of "the massive, two-day long discussion" (Wyczański, 1999, 163). The type of evidence and methods to gather and analyze the data as well as the central stage where the model was presented were inspired by and connected to the Parisian Annales and Braudel himself (Wyczański, 1999).

The far-reaching conclusion that Wyczański and Topolski drew in the course of their research indicated that international trade and grain export had a marginal role on the development of the early modern Polish economy. Their research results and interpretations, although influential in Poland, did not make it into mainstream economic history during the Cold War as the dominant Braudel paradigm favored the rival interpretations of Kula and Małowist. In an interview I conducted in 2002, Wyczański confirmed that the evidence he and Topolski provided devastated the colonial thesis of dependence on the grain trade and—on the optimistic eve of Poland's EU integration—was satisfied that his interpretation had not become part of the Marxist one. The third volume of Braudel's *Civilization and Capitalism* quotes his method of estimating the historical GDP even if his overall interpretation assigns Poland—against Wyczański's findings—the role of early periphery of the rising world economy (Braudel, 1992/3, 311).

Historians in Hostile Isolation: Halecki and Małowist (On the Two Sides of the Iron Curtain)

In the fall of 1956, during Wyczański's first stay at Parisian Annales school, he met Jerzy Giedroyć (Wyczański, 1999), a legendary editor of *Kultura* [*Culture*]. *Kultura* was the journal of literature, culture, and politics that was probably most seminal for the Polish intellectual alternative to the communist one that became truly hegemonic in the 1980s. Paris, a traditional site of Polish political and intellectual émigrés since the early nineteenth century, hosted after World War II the anti-communist Instytut Literacki in the Maisons-Laffitte; an important center of historically influenced political thought involving both professional and amateur historians in exile (Stobiecki, 2015). Animated by the strong personality and extended social networks of Jerzy Giedroyć, the periodicals *Kultura* and *Zeszyty Historyczne* [*Historical Notebooks*, hereafter *Zeszyty*], as well as Giedroyć's correspondence, Instytut Literacki [Literary Institute] influenced Polish historical studies and public debates in the 1980s and 1990s—the decades preceding and following the end of the Cold War.

Unlike Wyczański, other Polish economic historians who visited the Annales school did not establish contacts with Giedroyć's milieu. Kula neither participated in nor followed the *Zeszyty*'s debates (Kula, 2010). Kula's son Marcin does not even include Giedroyc or related entries in the index of his memoir. He briefly claims that they never visited Giedroyc. Kula found in his parents' letters "only two traces of that they read émigré journals and these are both references to the professional subject matters" (594), specifically to comments that *Teki Historyczne* [*Historical papers*] and *Kultura* published in 1958 and 1959 on Kula's book *Rozważania o historii* [*Reflections on History*] (Kula, 2010, 594).

Unlike Kula, Wyczański not only met Giedroyć but also took advantage of his networking and fundraising skills to have the Polish National Library equipped with its first microfilming machine. Both the encounter in Paris and Giedroyć's assistance to the major governmental institution was not only legal but also welcome in the relatively liberal atmosphere of the late 1950s. A historian of religion, Jerzy Kłoczowski, who cooperated with the Annales' medievalists, stayed in touch with institutions and scholars in exile, including Oskar Halecki, whose work he subsequently popularized among researchers of East Central Europe.

Biographies and works by Marian Małowist and Oskar Halecki illustrate how a common institutional background at the University of Warsaw in the interwar period and the same specialization in Polish medieval and early modern history were outweighed by the divergent social background, ideological beliefs, and professional experience on the two sides of the Iron Curtain. As a result, rival terminology and depictions of the region were created: the émigré historian Halecki's concept of East Central Europe, defined by its religious, geopolitical, and ethnic characteristics, and Małowist's concept of Eastern Europe, defined by its unfavorable patterns of economic exchange with Western Europe and the social institution of serfdom. The latter became popularized in Western historiography by Wallerstein while the former remained a domain of the American historiography of Peter F. Sugar, Donald W. Treadgold, and Piotr Wandycz and in Cold War Poland with historians associated with Catholic University in Lublin.

Both Małowist and Halecki are nowadays treated as classic historians in Poland. What is more, despite Halecki's emigration in 1945 and the forty-five-year ban on Halecki's research before 1989, his biography and writings have become an object of intensive research in the last 30 years (Cisek, 2009; Dąbrowska, 2012–2014; Kłoczowski, 1994).

Oskar Halecki (1891–1973) was born in Vienna as Oskar Ritter von Halecki, in an Austrio-Hungarian family, "where nobody spoke Polish and [where] German was a lingua franca." His ancestors included "representatives of nearly all nations of Central-Eastern Europe" (Kłoczowski, 1994, 397), including a Croatian *zupan* and a Polish king's secretary at the Brzesc conference that led to the Polish-Lithuanian Union of 1596 (Zadencka, 2015). Although born and raised in Vienna, he abandoned economic and social history studies at the University of Vienna and moved to the Jagiellonian University, the medieval Polish university located in Krakow, at that time the main center of Polish national movement in Austria-Hungary. In 1913, he defended his PhD dissertation on the religious history of the Polish-Lithuanian Commonwealth of the sixteenth century (Cisek, 2009).

In 1918–19, Halecki served as a member of the Polish delegation at the peace conference in Versailles as a historical expert, polyglot, and a member of a European elite (Wolff, 2020). Shortly after Poland was granted independence, the eager patriot accepted leadership in a newly created research committee on the history of Eastern Europe at the University of Warsaw. He became a professor at the University of Warsaw in 1919 and published *History of Jagiellonian Union* (1920) and *History of Relationships between Rome and Byzantium* (1930) (Cisek, 2009). In 1927, on the initiative of Marceli Handelsman and Halecki, the Federation of Historical Societies of Eastern Europe was founded, and in 1933 at the World Congress of Historians in Warsaw, a separate section devoted to the history of Eastern Europe was created (Troebst, 2011, 146).

In the University of Warsaw's history department, Małowist attended Halecki's classes but decided to write his MA thesis and PhD dissertation under the supervision of another medievalist, Marceli Handelsman. Why did Małowist not sign up for Halecki's seminar? Religious and political as well as social class divisions seem to have played a role in the tense atmosphere of the 1930s where a Jewish student from the working-class city of Łódź with no prospects for employment at the university could not find common ground with a well-traveled, diplomatically experienced professor who was a devoted although never anti-Semitic Catholic activist (Siewierski, 2016, email).

Halecki's experience of World War II was very different from Małowist's life in the Warsaw ghetto, extermination of his family, and hiding in the Polish countryside. The introduction of the Soviet-style regime had the opposite effect on their professional fates. Since the outburst of the World War II in 1939 found Halecki on a trip to Switzerland, he did not return to Poland and escaped Nazi occupation in the United States. Although he lectured during the Cold War in Paris, Montreal, Brugia, Freiburg, and Rome (Cisek, 2009) and kept in touch with Polish political and intellectual institutions in exile in Paris, London, and Rome, New York City became his home from August 1940. Halecki was a crucial figure in organizing Polish academic institutions in the United States including the Polish Institute of Arts and Sciences in America and the Polish-American Historical Association. One of the few Polish historians with international recognition in the *Encyclopedia of Historians and Historical Writing* (1999; Stobiecki, 2014), Halecki participated in international historical congresses and published extensively in English. As a professor at Fordham University in New York City from 1944 to 1961, he became a part of the U.S. academic world, even if not of its mainstream. He worked as board member of *Slavic Review* and the *Journal of Central European Affairs*.

As an émigré in America, Halecki continued to work as a historian and assumed the role of lobbyist for the region, which from 1943 he termed East Central Europe (Halecki, 1943, 1944). His double engagement—academic and political—pushed him to treat history writing as "a basis for a political strategy aimed at breaking down the Russian/ Soviet Empire or at least at defeating Russian supremacy in East-Central Europe" (Filipowicz, 2015, 259). His speeches and publications aimed to combine the history of Europe's borders and divisions with current geopolitical solutions. "During the dramatic war years he spoke on behalf of the small countries between Germany and Russia, closely connecting historical reflections with a vision of the future of the region" (Kłoczowski, 1995, 12). Halecki's articles written during the war (1943, 1944) encouraged the Western allies to defend the borderlands of Western civilization, while *The Limits* published in 1950 tackled the post-Yalta order and its predecessor Nazi imperialism, which threw the region from one occupation to another. These as well as *The Borderlands* (1952) were addressed to Western audiences and sought to undermine the image of communist bloc countries as an eternal and natural sphere of Russian influence.

The region of Halecki's attention was meant to convey the historical links with the two imperial powers—Germany in Central Europe and Russia in Eastern Europe—but the same time emphasize the region's separateness from both. In sketching the borders of the region, Halecki applied religious, political, historiographical, and ethnic criteria. It was the marginal role of Protestantism that provided the concept of East Central Europe with cohesiveness. An expert on Christian history and doctrine, he was sure that "from the point of view of religious doctrine and even of ecclesiastical organization, the differences between Protestantism and both Catholicsm and Orthodoxy are much greater than those which separate the two latter" (Halecki, 1950, 106). Among the political factors behind the divisions of Europe, Halecki attached greatest importance to the international strength and size of political organisms. Halecki's East Central Europe would have been an area of small nations, many but not all of them ethnically Slavic, with weak states or stateless throughout modern history, and therefore, neglected in world historiography. Halecki was a devoted Catholic and Catholic activist before emigration. His work in the American academic world and lobbying against the atheist communist government established in Poland in 1945 seem to have strengthened that identity (Filipowicz, 2015). Małowist and Halecki had a chance to meet at the postwar international historical congresses—the former as a major participant of Poland's national delegation and the latter as a member of the marginal Slavic studies group. As much as historians in exile, with Halecki as a leader, invested in the principle of national delegations to the congresses held after 1918 to gain recognition for the historiographies of the Central East European countries created in Versailles, after World War II he sought recognition for émigrés as Polish historians although not a part of the communist Poland's delegation (Zadencka, 2015, 165). In the 1950 congress in Paris, the 1955 congress in Rome, and the 1960 congress in Stockholm, Polish exile historians participated via the International Commission of Slavic Studies whose panels and publications were left out of the main congress discussions and publications. From the 1965 congress in Vienna, they were incorporated into the representations of their new countries where they did not belong to the mainstream either (Zadencka, 2015, 171–172).

Yet, probably the most profound obstacle that émigré historians struggled with came from the shift in the intellectual paradigm of how historical studies were being practiced after the World War II. Methodologically, Halecki, like other anti-communist historians in exile, followed "individualistic historicism" rooted in the German philosophical and history writing tradition. Its main trademarks included search for the spiritual meaning and goal in history and concentration on individuals who bear this meaning (Mękarski, 2015, 355). Filipowicz (2015) and Mękarski (2015) stress that even the most successful exile historians were surprised by the degree to which postwar historiography shifted toward the social and economic history of large social groups and long social processes and away from the political history of exceptional individuals that framed their approach. They observed with astonishment the increasing influence of Marxism and the Annales school on mainstream historiography in international congresses; at U.S., British, and French universities; and in the publication market. Marian Kukiel, a leader of London historians in exile, when confronted with the influence of the Annales school at the first postwar International Congress of Historical Sciences, "went so far as to say that the Congress held in Paris in 1950 could be regarded as marking the beginning of the crisis of Western historiography. (…)

The stress on the material dimension of life (...) seemed to deflect attention from the spiritual sphere of reality which in the eyes of most émigré authors appeared to be much more important" (Mękarski, 2015, 355).

Zadencka notes the clash between Halecki and Małowist at the Paris congress in 1950. Halecki commented critically on Malowist's report on the state of the art of historical research in Poland. Halecki criticized the use of the term "social class" (Zadencka, 2015, 172) in Małowist's speech. A couple of years later, and with reference to Małowist-led medieval scholarship, Halecki diagnosed Poland's historiography with negligence for ignoring the role of Christianity and the Catholic Church in the development of the region in the late Middle Ages (ibid., 174). Nor did he agree with the Polish delegation's depiction of the Polish-Lithuanian nobility as a destructive social and political actor (Zadencka, 2015, 174).

Yet, there are also testimonies to Halecki's account of Małowist as one of the few creative and solid researchers in communist Poland (Zadencka, 2015, 184). Halecki wrote a welcoming review of Małowist's *Kaffa* (Halecki, 1948; Siewierski, 2016, 103). This short note might be read as Halecki's postwar willingness to develop an academic exchange across the Iron Curtain that was not picked up by the Polish colleagues. The recognition was not mutual. Małowist never referred to Halecki's assertion of the positive role of Christianity and nobility in Poland's development. A declared atheist, religion-blind in his medieval studies, he treated Catholicism and its political manifestations in Poland with the highest suspicion. A supporter of the anti-communist opposition since the 1970s and the Solidarity movement in 1980s, he was concerned with the alliance that the movement established with Catholic Church, as it "in his eyes – carried very conservative, anti-Semitic content" (Siewierski, 2016, 76–77). His works remain silent about the role of religion or culture in economic development while he presented the Polish nobility of the sixteenth century as disloyal to the lower classes and taking advantage of the country's growing dependence on grain exports to the West (Małowist et al., 2000).

Conclusion

The Parisian Annales, Prato seminars, and IEHA in the era of charismatic researcher and efficient academic manager Fernand Braudel played a crucial role in establishing academic cooperation across the Iron Curtain.

They constituted the main institutional framework, the center of circulation of ideas, and a guard of the paradigm of history writing that allowed Polish historians' research results and bold historical interpretations to enter the international debate on the origins of capitalism and the rise of the West. The works of Witold Kula and Marian Małowist, convinced Marxists but relatively independent from Party orthodoxy, impacted the course of the debate and the image of Eastern Europe's past. Via Braudel's and Wallerstein's works and the Brenner debate, the region entered the debate as an early grain-export-dependent periphery of the capitalist world-system in the sixteenth century. Along with academic excellence, this international influence was a combination of post-Yalta circumstances that favored economic history in the Academy above other sub-disciplines with the postwar shift in the Western history research paradigm that made the financially unequal exchange between East European and Western scholars intellectually and prestigiously beneficial in the orbit of Braudel-dominated institutions.

The modeling and large interpretations that became standard in the works of economic historians in the 1960s and 1970s (Kula, Wyczański, and Topolski) resulted, according to the testimonies of historians themselves, from the influence of the Annales, American cliometrics, and Anglo-Saxon economic studies on developing economies. Mentality, as a subject of historical change and a mode of the mind that connects economy and culture via consumption, became from the 1960s an important component of depicting and explaining the past starting with the works of Topolski and Wyczański.

The ideological and methodological differences between the Polish historians at home and in exile, as well as "Braudelization" of postwar history research resulted in isolation of the rich network of historical institutions and personalities in exile from their colleagues in Poland and increasingly from the mainstream history writing of which the latter was a part.

References

Assorodobraj-Kula, N. (1963). "Żywa historia": świadomość historyczna: Symptomy i propozycje badawcze. *Studia Socjologiczne*, *2*(9), 5–45.

Bębnowski, D. (2019). Sprawozdanie z międzynarodowego seminarium naukowego "Behind the Iron Curtain", Praga, 1–2 III 2019 R. *Roczniki Dziejów Społecznych i Gospodarczych*, *80*, 433–436.

Behr, V. (2022). *Powojenna historiografia polska jako pole walki: studium z socjologii wiedzy i polityki*. Wydawnictwa UW, in print.

Berg, M. (2015). East-West Dialogues: Economic Historians, the Cold War and Détente. *Le Mouvement Social, 259*(2), 33–58.

Bieżuńska-Małowist, I., & Małowist, M. (1987). *Niewolnictwo*. Czytelnik.

Braudel, F. (1971). *Historia i trwanie* with introduction by W. Kula and B. Geremek.

Braudel, F. (1972). *The Mediterranean and the Mediterranean World in the Age of Philip II*. Harper & Row.

Braudel, F. (1992). *Civilization and capitalism, 15th-18th Century* (vol. 1–3). University of California Press.

Brzezińska, A. (2021). French historiography and Oskar Halecki. *Historyka: Studies in Historical Methods, 51*, 93–105.

Cerman, M. (2012). *Villagers and Lords in Eastern Europe, 1300–1800*. Palgrave Macmillan.

Cisek, J. (2009). *Wstęp in: Oskar Halecki. Historyk – szermierz wolności*. IPN.

Connelly, J. (2000). *Captive University: The Sovietization of East German, Czech and Polish Higher Education 1945–1956*. University of North Carolina Press.

Dąbrowska, M. (Ed.). (2012–2014). *Oskar Halecki i jego wizja Europy* [Oskar Halecki and His Vision of Europe]. Uniwersytet Łódzki.

Filipowicz, M. (2015). Polish Historiography in Exile: On Selected Works and Ideas of Oskar Halecki, Henryk Paszkiewicz and Marian Kukiel. In M. Zadencka, A. Plakans, & A. Lawaty (Eds.), *East and Central European History Writing in Exile 1939–1989* (pp. 249–262). Brill.

Górny, M. (2013) .*The Nation Should Come First : Marxism and Historiography in East Central Europe* (A. Górny, Trans.). Peter Lang.

Górny, A. (2021). Writing History in Twentieth-century Eastern Europe. In W. Borodziej, F. Laczo, & J. Puttkamer (Eds.), *The Routledge History Handbook of Central and Eastern Europe in the Twentieth Century* (Vol. 3, pp. 194–255). Taylor & Francis.

Guzowski, P., Kuklo, C., Poniat, R., Mikulski, K., & Tęgowski, J. (2015). *O historii integralnej w ujęciu Andrzeja Wyczańskiego*. Instytut Badań nad Dziedzictwem Kulturowym Europy.

Halecki, O. (1943). East Central Europe in Postwar Organization. *Annals of the American Academy of Political and Social Science, 228*, 52–59.

Halecki, O. (1944). The Historical Role of Central-Eastern Europe. *Annals of the American Academy of Political and Social Science, 232*, 9–18.

Halecki, O. (1948). Review: Marian Małowist, Kaffa – kolonia genueńska na Krymie. *Speculum: A Journal of Mediaeval Studies, 4*, 717–720.

Halecki, O. (1950). *The Limits and Divisions of European History*. Sheed & Ward.

Halecki, O. (1952). *Borderlands of Western Civilization: A History of East Central Europe*. Ronald Press Company.

Kamecka, M. (2004). Ludzie i instytucje. O współpracy historyków polskich i francuskich. In C. Kuklo, & P. Guzowski (Eds.), *Cała historia to dzieje ludzi: studia z historii społecznej ofiarowane profesorowi Andrzejowi Wyczańskiemu w 80-tą rocznicę urodzin i 55-lecie pracy naukowej. Studia z historii społecznej* (pp. 407–421). Wydaw. Uniwersytetu w Białymstoku.

Kilias, J. (2021). Not Only Scholarships: The Ford Foundation, Its Material Support, and the Rise of Social Research in Poland. *Serendipities. Journal for the Sociology and History of the Social Sciences, 5*(1–2). https://doi.org/10.7146/serendipities.v5i1-2.126089

Kłoczowski, J. (1994). Oskar Halecki i jego walka o miejsce Polski w Europie. In H. Bułhak (Ed.), *Z dziejów polityki dyplomacji. Studia poświęcone pamięci Edwarda hr. Raczyńskiego, Prezydenta Rzeczypospolitej Polskiej na wychodźstwie* (pp. 397–406). Wydawnictwa Sejmowe.

Kłoczowski, J. (1995). *East Central Europe in the Historiography of the Countries of the Region*. Institute of East Central Europe. https://biblioteka.teatrnn.pl/Content/10987/East_Central_Europe.pdf

Kochanowicz, J. (2016). A Periphery at the Center of Attention: Economic History in Poland. In F. Bodizzoni & P. Hudson (Eds.), *Routledge Handbook of Global Economic History* (pp. 195–213). Routledge.

Kochanowicz, J., & Sosnowska, A. (2011). Economic History of Pre-industrial Poland: An Obsolete Subject? In *Dove va la storia economica? Metodi e prospettive: secc. XIII-XVIII = Where Is Economic History Going? Methods and Prospects from the 13th to the 18th Centuries*. Firenze University Press.

Kormanowa, Ż. (1951). Referat podsekcji historii sekcji nauk społecznych i humanistycznych na Kongresie Nauki Polskiej. *Kwartalnik Historyczny, 2*, 253–325.

Kula, W. (1956). *Szkice o manufakturach w Polsce XVIII wieku*. PWN.

Kula, W. (1976). *An Economic Theory of the Feudal System: Toward a Model of the Polish Economy 1500–1800*. New Left Books.

Kula, W. (1996). *Rozdziałki*. Trio.

Kula, W. (2001). *The Problems and Methods of Economic History*. Ashgate.

Kula, M. (2010). *Mimo wszystko bliżej Paryża niż Moskwy. Książka o Francji, PRL i o nas, historykach*. WUW.

Kula, W., & Leskiewiczowa, J. (Eds.). (1970–1982). *Społeczeństwo polskie XVIII i XIX wieku. Studia o uwarstwieniu i ruchliwości społecznej* [*Polish Society in the 18th and 19th c. Studies in Social Stratification and Mobility*] (vol. 4–7). PWN.

Łepkowski, T. (1990). Wątki trzecioświatowe w twórczości Witolda Kuli. In M. Kula (Ed.), *Dziedzctwo Witolda Kuli*. Uniwersytet Warszawski.

Mączak, A., & Tygielski, W. (2000). *Latem w Tocznabieli*. Biblioteka "Więzi".

Małowist, M. (1958). Poland, Russia and Western Trade in the Fifteenth and Sixteenth Centuries. *Past and Present, 13*, 26–41.

Małowist, M. (1959). The Economic and Social Development of the Baltic Countries 15–17th centuries. *The Economic History Review, 12*, 177–189.

Małowist, M. (1960). A Certain Trade Technique in the Baltic Countries in the 15th to 17th Centuries. In *Poland at the 11th International Congress of Historical Sciences* (pp. 103–116). Instytut Historii PAN.

Małowist, M. (1966). The Problem of the Inequality of Economic Development in Europe in the Later Middle Ages. *The Economic History Review, 19*, 15–28.

Małowist, M. (1967). Review of: Civilisation materielle, economie et capitalism, XV-XVIII siècle by F. Braudel. *Przegląd Historyczny, 58*, 691–696.

Małowist, M. (1969). *Europa a Afryka Zachodnia w dobie wczesnej ekspansji kolonialnej* [*Europe and West Africa in the Era of Early Colonial Expansion*]. PWN.

Małowist, M. (1974). Problems of the Growth of the National Economy of Central-Eastern Europe in the Late Middle Ages and the Beginning of the Sixteenth Century. *The Journal of European Economic History, 3*, 319–357.

Małowist, M. (1992). *Konkwistadorzy portugalscy*. Państwowy Instytut Wydawniczy.

Małowist, M. (2006). *Wschód a Zachód Europy w XIII–XVI wieku: konfrontacja struktur społeczno-gospodarczych*. Wydawnictwo Naukowe PWN.

Małowist, M. et al. (2000). Marian Małowist on History and Historians. In A. Polonsky (Ed.), *Polin: Studies in Polish Jewry. Volume 13: Focusing on the Holocaust and Its Aftermath* (pp. 331–344). Liverpool University Press.

Małowist, M., Batou, J., & Szlajfer, H. (2010). *Western Europe, Eastern Europe and World Development, 13th-18th: Collection of Essays of Marian Małowist*. Brill.

Małowist, M., & Clarke, H. W. (1966). *The Social and Economic Stability of the Western Sudan in the Middle Ages*. Corpus Christi College.

Mękarski, A. (2015). In Whose Name Is the Story Told? The Émigré Critique of Method in the Historiography in the Polish People's Republic. In M. Zadencka, A. Plakans, & A. Lawaty (Eds.), *East and Central European History Writing in Exile 1939–1989* (pp. 343–359). Brill.

Norkus, Z. (2012). Modeling in Historical Research Practice and Methodology: Contributions from Poland. *History and Theory: Studies in the Philosophy of History, 51*(2), 292–304.

Pleskot, P. (2010). *Intelektualni sąsiedzi: kontakty historyków polskich ze środowiskiem "Annales" 1945–1989*. IPN.

Pomian, K. (1978). The Impact of the Annales School in Eastern Europe. *Review, 3/4*, 101–118.

Romek, Z. (2010). *Cenzura a nauka historyczna w Polsce 1944–1970*. Wydawnictwo Neriton, Instytut Historii PAN.

Rutkowski, T. P. (2007). *Nauki historyczne w Polsce 1944–1970*. Uniwersytetu Warszawskiego.

Siewierski, T. (2016). *Marian Małowist i Krąg Jego Uczniów: Z Dziejów Historiografii Gospodarczej w Polsce*. Oficyna Wydawnicza.

Sosnowska, A. (2019). *Explaining Economic Backwardnes: Post-1945 Polish Historians on Eastern Europe*. Central European University Press.

Stobiecki, R. (2005). *Klio na wygnaniu. Z dziejów polskiej historiografii na uchodźctwie w Wielkiej Brytanii po 1945 r.* Wydawnictwo Poznańskie.

Stobiecki, R. (2014). *Historycy polscy wobec wyzwań XX wieku*. Wydawnictwo Nauka i innowacje.

Stobiecki, R. (2015). Polish Exile Periodicals as a Dialogue Forum: Teki Historyczne, Polish Review, Zeszyty Historyczne. In M. Zadencka, A. Plakans, & A. Lawaty (Eds.), *East and Central European History Writing in Exile 1939–1989* (pp. 282–296). Brill.

Stobiecki, R. (2020). *Historiografia PRL*. Wydawnictwo Uniwersytetu Łódzkiego.

Topolski, J. (1974a). The Manorial-Serf Economy in Central and Eastern Europe in the 16th and 17th Centuries. *Agricultural History, 48*(3), 341–352.

Topolski, J. (1974b). Economic Decline in Poland from the Sixteenth to the Eighteenth Centuries. In P. Marle (Ed.), *Essays in European Economic History 1500–1800*. Clarendon Press.

Topolski, J. (1976). *Methodology of History* (O. A. Wojtasiewicz (Ed.)). Polish Scientific Publishers.

Topolski, J. (1977). *Gospodarka polska a europejska w XVI-XVIII wieku*. Wyd. Poznańskie.

Topolski, J. (1982). *Prawda i model w historiografii*. Wydawnictwo Łodzkie.

Topolski, J. (1983). Teoria wiedzy historycznej. Wydawnictwo Poznańskie.

Troebst, S. (2011). Halecki Revisited: Europe's Conflicting Cultures of Remembrance. In P. Meusburger, M. Heffernan, & E. Wunder (Eds.), *Cultural Memories: Knowledge and Space*. Springer.

Troebst, S. (2021). From Halecki to Hann: The Historiography of Historical Regions. In D. Kaneff & K. W. Endres (Eds.), *Explorations in Economic Anthropology: Key Issues and Critical Reflections*. Berghahn.

Wallerstein, I. (2011). *The Modern World-System I: Capitalist Agriculture and the Origins of the European World-Economy in the Sixteenth Century*. University of California Press.

Wierzbicki, A. (2004). W stronę postaw otwartych. 'Annales' w refleksji historyków polskich po II wojnie światowej. *Klio Polska. Studia i materiały z dziejów historiografii polskiej po II wojnie światowej* (pp. 49–80).

Wolff, L. (2020). *Woodrow Wilson and the Reimagining of Eastern Europe*. Stanford University Press.

Wyczański, A. (1960). *Studia nad folwarkiem szlacheckim w Polsce w latach 1500–1580*. PWN.

Wyczański, A. (1964). *Studia nad gospodarką starostwa korczyńskiego 1500–1660*. PWN.

Wyczański, A. (1965). *Polska - Rzeczą Pospolitą Szlachecką 1454–1764*. PWN.

Wyczański, A. (1970). *Polska w epoce odrodzenia: państwo, społeczeństwo, kultura*. Wiedza Powszechna.

Wyczański, A. (1987). *Dogonić Europę, czyli Polska w czasach Zygmunta I (1506–1548)*. Krajowa Agencja Wydawnicza.

Wyczański, A. (1993). Odnowienie stosunków naukowych z Francją w 1956 roku. *Kwartalnik Historyczny, 100*(4), 265–268.

Wyczański, A. (1999). Moja współpraca naukowa z Francją. *Nauka, 6*(4), 161–167.

Wyczański, A., Flandrin, J.-L., & Tollet, D. (1985). *La consommation alimentaire en Pologne aux XVIe et XVIIe siècles*. Publications de la Sorbonne.

Zadencka, M. (2015). Polish Exile Historians at the International Historical Congresses. In M. Zadencka, A. Plakans, & A. Lawaty (Eds.), *East and Central European History Writing in Exile 1939–1989* (pp. 156–189). Brill.

CHAPTER 7

Yugoslavia: Economic Historiography Between National and International Context

Žarko Lazarević

After World War II economic historiography in Yugoslavia shared the fate of historiography in general. In the new ideological and political reality, historiography was assigned a central place in the ideological apparatus. Historiography was not only to become Marxist, but also to support the new political regime and direct its studies toward the people and their role in historical processes, as well as toward the Communist Party of Yugoslavia (hereafter the Party) as the people's representative. After the Cominform dispute and the gradual liberalization of the regime, direct political intervention in science and culture became less intense and was partially abandoned. But not in the field of history. Throughout the existence of the communist regime, historiography was under political and ideological control—first directly and then indirectly. The history commissions, which operated within the framework of the

Ž. Lazarević (✉)
Department of Economic and Social History, Institute of Contemporary History, Ljubljana, Slovenia
e-mail: zarko.lazarevic@guest.arnes.si

A. Dolezalova and C. Albrecht (eds.), *Behind the Iron Curtain*, Palgrave Studies in Economic History,
https://doi.org/10.1007/978-3-031-31578-7_7

central organs of the Party, existed until the disintegration of the Yugoslav state. They organized many debates on modern historiographical tendencies, thus indirectly influencing research efforts and setting the limits of interpretative freedom (Režek, 2014, 986).

Even so, the nation had already completely replaced class as the central theme of historiography by the mid-1960s (Brunnbauer, 2003, 435). There were no obstacles to the inclusion of class or revolutionary traditions in the concept of nation and national history (Marković, 2004, 51). The synergy between the principles of "class" and "nation" in historiography was already evident during the establishment of communist authority. Historical materialism, with its emphasis on social and economic history, enabled even small nations without a state to attain the status of a "historical nation" (Grafenauer, 1947, 22–26), which was an important focus for the Yugoslav nations. The nation or state remained the central framework of socialization and an important factor in the historical process (Brunnbauer, 2003, 435). Once the nation had completely replaced class as the frame of reference for research, the diversity of historical experiences, and thus of interpretations, became apparent. By the 1980s, debates among historians foreshadowed the political divisions that would lead to the end of Yugoslavia (Banac, 1992, 1084–1104; 1995, 39–65; Repe, 1996, 69–78; 1999, 312–325).

Yugoslav historiography functioned as a collection of regional (republican and national) historiographies (Stallaerts, 1999, 315–336). This process was closely related to the policy of decentralization, in which powers were transferred from the federation to the republic level. Decentralization strengthened the historiographies of the individual Yugoslav nations (Brunnbauer, 2003, 435). Yugoslav historiography was also heterogeneous in the field of methodology. Different orientations as well as different levels of professional standards prevailed (Stallaerts, 1999, 330). A positivist-oriented history was typical. The focus on events and "facts" that had to be substantiated by written sources enabled historians to conduct their research during the period when ideological pressures were strongest (Brunnbauer, 2003, 439). At the same time, a minority of historians were strongly influenced by international trends, the French Annales school, or the concepts of social history. However, some historians insisted on Marxist historical determinism (Najbar Agičič, 2013, 274–300; Repe, 1996, 72; 1999, 316–317).

Economic historiography developed within individual republics, based on the ethnic principle, except for Bosnia and Herzegovina. The republics with several institutions (universities and institutes) and a longer tradition of research were at the forefront of economic historical studies. Therefore, Croatian and Serbian economic historiography was leading. By contrast, historiography, including economic historiography, was only beginning to emerge in Bosnia and Herzegovina, Montenegro, and Macedonia. Compared to the community of historians, the number of economic historians was modest.

In the beginning of the 1950s, economic historiography was still in its infancy. It faced the challenges of asserting itself as an independent discipline. While the economic history of the Middle Ages was already established, the study of more recent times was completely neglected. The new rulers were interested precisely in modern history—the history of the capitalist economy. Interestingly, a literary version of the communist interpretation of pre-World War II economic history emerged. In the second half of 1945, a novel by the well-known Slovenian writer Lovro Kuhar was published with the title *Jamnica* (Prežihov, 1945). As a valley, Jamnica was a symbol of Slovenia, but with a little imagination the novel could also be read in a Yugoslav context. In the 1930s, the author was one of the leading Yugoslav communists (Gabrič, 2010). In the novel, Kuhar deals with the economic and social issues of the broad sections of the population in the grip of advancing capitalism, industrialization, the emergence of the working class, class struggles, economic crises, the impoverishment of the population, and the penetration of capitalist logic into rural communities. The basic contours of understanding the economic past were thus already established in this novel. At the same time, it raised research questions for future economic historians.

The 1960s and 1970s marked a structural change in the field of economic history in Yugoslavia. In the formative period, economists who also held Marxist views were at the forefront. Thereafter, historians took the main role in economic historiography. The Marxist approach gradually faded away—not as a public declaration, but in the practice of economic historians. Reference to Marxist concepts of social analysis among historians became increasingly rare. Capitalism was replaced by the concepts of industrialization and, later, economic modernization. In this respect, economic historians did not abide by the wishes of the authorities. They devoted themselves to various aspects of economic development and shed light on the local phenomena that resulted from global processes, social structures, and restructuring. They also focused on the masses as the "driving force of history" as well as on the different social classes (workers, peasants, students, civil servants, and so on). In methodological terms, economic historiography followed the traditional historical (positivist) pattern with an emphasis on empirical evidence, descriptive analysis, and the partial incorporation of quantitative methods.

This chapter focuses on the development of economic historiography, methodological issues, changes in scientific policy depending on political reforms, and international cooperation in the field of science, with an emphasis on the community of economic historians.

Scientific Policy and Institutions

In 1968, the journal *Nature* published an article informing the international public about the organization, politics, and structure of scientific activities in Yugoslavia "The changes in Yugoslav scientific policy in the postwar period reflect the changing social structure of the country" (Malsimović, 1968, 846–849). Practices depended on the reform processes in the communist regime, while international cooperation was determined by the international political orientation of the country. After World War II, science policy went through several developmental periods. Each of these periods brought changes in science policy and decentralization of science as responsibilities were transferred from the federation to the individual republics. These periods correspond to the Yugoslav social development models (Ramnath, 1988, 2052–2054). As the country was liberalized so was science policy and international cooperation. Leaving the socialist bloc in 1948, Yugoslavia was gradually integrated into the Western political, economic, and scientific framework.

Yugoslavia could not escape the fate of a country on the periphery of European modernization processes. Before World War II, scientific work was carried out at the universities and academies of sciences. In total, the scientific community at the three universities and academies of sciences in Zagreb, Belgrade, and Ljubljana numbered about 1,000 people (Malsimović, 1968, 846). This was far too few for the plans of the communist authorities, for education and science were to be the focal points of social transformation. Extensive investments in the scientific infrastructure began, so that the first years after World War II became the formative period for scientific institutions. The transformation of Yugoslavia into a communist country depended on the Soviet Union, both politically and economically. Science was also subjected to Sovietization. In the early stages, science policy was centralized at the federal level. The authorities made no secret of their ideological approach, and Marxist ideology was to be the basis for scientific work. Scientific activities were also subject to political control.

In the new society, science had to be in the "service of the people," and the goal of this service was determined by the Party. The universities were to become "cadre schools" for the new socialist intelligentsia, which was to work based on the Marxist worldview. By the early 1950s, every republic except Montenegro had its own university. In the 1970s, the university network expanded. Within the framework of the academies of arts and sciences, institutes for various branches of science were established on the Soviet model. The Yugoslav Academy of Sciences and Arts in Zagreb, which had the longest tradition in the country, had 37 such institutes in the early 1950s. Twenty-five institutes operated under the auspices of the Serbian Academy in Belgrade and twelve under the Slovenian Academy in Ljubljana. The institutes established within the academies were intended to introduce young Marxist researchers to science and to promote technological development to support and facilitate the country's economic transformation into an industrial society. Among them were historical institutes, but not institutes of economic history. The social sciences and humanities had the task of improving the general education of the population as much as possible and of laying the "scientific" foundations for the socialist ideological transformation of society, culture, education, and science itself. The Soviet Union was expected to provide material, technical, and professional support in building the system of scientific institutions. Soviet scientific achievements constituted the main reference. Yugoslav universities adopted Soviet curricula and study literature. For some years, most of the books supplied were in Russian. Science was divided into a Soviet and a bourgeois variety, with only Soviet science ascribed positive political qualities and social legitimacy (Bondžić, 2018, 206; Gabrič, 2008, 305–309).

A gradual shift away from Soviet practices began after 1948. The withdrawal of aid and the cancelation of training of specialists in Soviet universities and institutes had serious negative effects on the Yugoslav scientific system. The first reaction, when relations with Western countries had not yet been established, was to try to become self-sufficient. All efforts were directed at training the needed specialists at home, and the scientific institutes were to increase their research efforts without Soviet help (Bondžić, 2018, 207). Exclusion from the community of communist countries led to the establishment of a different model of society.

From the 1950s onward, the country's reform processes were characterized by decentralization, that is, the shift of development policymaking from the federal level to the individual republics and, in some cases, to individual institutions (Ramanth, 1988, 2052–2054). During the second development model, the network of scientific and higher education institutions continued to grow. As Yugoslavia gradually emerged from political isolation and established itself as an intermediate link in the Cold War dichotomy between East and West, it began to open to the West in the scientific sphere as well. Cooperation with Western institutes and universities began to expand, as did cooperation with the countries of the so-called Third World in Asia and Africa from the mid-1950s.

Federal laws established only the common framework, while the main thrusts of science and higher education policy were implemented at the republic level. In 1957 a new law on scientific activity was passed, completely abolishing the Soviet organization of science. In addition, in 1965 a law was passed on the coordination of scientific work at the federal level, and a federal fund was established to finance science. Thus, the organization of international scientific cooperation remained at the federal level. The Federal Fund for the Financing of Science was to provide funds for nationwide scientific projects involving institutions from different parts of the country. The individual republics took over the financing of their own scientific and higher education institutions. Decentralization stimulated the further development of those institutions.

This period can also be considered the end of the formative period. The results were remarkable. According to the official records, in 1956 there were 730 scientific institutions (institutes, clinics, laboratories, and so on) in the country. Before World War II, there had been 215. Between 1953 and 1956 alone, the number of researchers at the universities increased by 27 percent and at the various institutes by as much as 55 percent. Not only the institutional expansion of the system but also the expansion of research activities at existing institutions drove that growth (Bondžić, 2018, 206–212; Gabrič, 2008, 309–313). In the following decades, the number of researchers continued to increase. In 1965 there were 6,511 researchers, in 1980 23,164, and in 1987 32,633. Most researchers were employed in institutes, while a small percentage worked in the development departments of various companies. Overall, there were 9.6 researchers per 10,000 inhabitants and 39.4 researchers per 10,000 employees in Yugoslavia in 1978. Analyses have shown that the gap in staffing was relatively smaller compared with other countries, while the gap in the volume of investment was more pronounced. The growing number of researchers is evidence of the governments' awareness of the need to invest more in science and research. At the level of policy statements and economic and social development strategies, the federal and republican governments attached great importance to the promotion of innovation and scientific research as a fundamental instrument of development policy. The profound economic crisis, inflation on the way to hyperinflation, and the erosion of social cohesion in the country in the 1980s created a situation in which investment in science and research stalled (Bondžić, 2018, 215–217).

By the early 1960s, due to the international opening, freedom to travel, and the beginning of mass migration to the Western countries, Yugoslavia faced an emigration of experts in the field of science. The first cases were observed as early as the 1950s, when people did not return from stays abroad or took up new jobs during their stays and subsequently emigrated. The first reaction of the authorities was to try to convince host countries to prevent the admission of Yugoslav scientists (Bondžić, 2011, 157–159). In the 1960s, the problems became even worse, as graduates emigrated abroad without looking for jobs at home. The emigration of Professor Anton Peterlin (1908–1993), a renowned Slovenian physicist who had earned his doctorate at Humboldt University in Berlin before World War II, caused a stir. Among the reasons he gave for his departure were inadequate funding for science, unsuitable conditions for in-depth research, and an uninspiring environment that did not sufficiently reward scientific excellence (Gabrič, 2008, 326–329). Another prominent case was Professor Jadran Ferluga (1920–2004), who had studied in Paris during the first two postwar years and completed his studies and dissertation at the University of Belgrade. He became a distinguished professor of the history of Byzantium and the mediaeval history of the Yugoslav lands and one of the most renowned Byzantologists in Europe. In 1970, he left the University of Belgrade to take the chair of Byzantine Studies at the University of Muenster in West Germany (Voje, 2004, 207–209).

The first migrants who moved to West European countries created networks that enabled chain migration. Especially problematic became the emigration of professionals who, with the help of international academic networks, went in search of better working conditions and much higher salaries (Dragišić, 2019, 99–116). In the 1960s and 1970s, when emigration from Yugoslavia reached its peak, most emigrants were medical doctors, stomatologists, engineers, technologists of various specialties, physicists, mathematicians, as well as some social scientists. Most of them moved to the countries of Western Europe, but many also emigrated to the United States, where 1,370 specialists moved in the ten years from 1963 to 1973 (Lukšič-Hacin, 2020, 113–128). By the end of the 1970s, 269 renowned engineers, 79 university professors, and 103 researchers had gone to the United States alone. Overall, it is estimated that between the 1960s and the late 1980s some 50,000 persons with higher education emigrated from Yugoslavia (Šoljan, 1991, 148–149).

The establishment of institutes for the history of the workers' movement in the late 1950s in all Yugoslav republics made an important contribution to strengthening (economic) history research (Gabrič, 2009, 17–48). Based on historical materialism, the institutes were to advance the study of the development of the labor movement, communist party, and economic and social development. These were subjects that, according to the communist authorities, had been neglected in historical research. It was not possible to talk about the working class without at the same time presenting economic and social conditions. Thus, the course of industrialization and economic development became a central theme (Kresal, 1973, 313–317). In the long run, the new institutes did not change the existing situation in the writing of history. While a small part of their attention was devoted to the history of the Party, most of the research took place in the direction of national political, cultural, social, and economic history and followed established historical methods (Gabrič, 2009, 17–48).

As institutions developed, so did the community of economic historians. The Yugoslav Association of Historians established, among other commissions, the Commission for Economic History in the 1960s, as did the historical associations in each republic (Mihelič, 2003, 577–578). The Commission for Economic History was a space for dialogue among economic historians, enabling the exchange of ideas and experiences. The commission had three main tasks: Bringing together economic historians in the country through conferences and other forms of cooperation; publishing the journal *Acta historico-oeconomica Iugoslaviae*; and participating in the preparations for the World Congresses of Economic History organized by the International Economic History Association or in other activities. The Commission for Economic History organized five conferences.

The first, held in 1972, dealt with the Industrial Revolution. The theme of the second conference in 1977 was transport. In 1980, the specifics of economic development and social structures in the various regions of Yugoslavia were discussed. In 1983, the focus was on forests and forestry at the interface between natural heritage and economic activities. In 1987, the meetings were concluded with lectures on livestock and its role in the economy (Cvetković & Matković, 2021, 199–202). The first foreign guest—the German agricultural historian Willi A. Boelcke—was invited to the last conference. The papers presented at the conferences were published in the journal *Acta historico-oeconomica Iugoslaviae*, which first appeared in 1974. It was the only journal in the country that dealt exclusively with economic history topics. The editorial board was based in Zagreb, which is not surprising since Croatian historiography was at the forefront of economic history research. The members of the editorial board came from all Yugoslav republics, and the editor-in-chief was the renowned Croatian economic historian Ivan Erceg (1919–2017), who was also, among other things, a professor of economic history at the Faculty of Economics at the University of Zagreb. For a short time, he also studied and specialized abroad and was especially influenced by his specialization in agricultural history at the University of Hohenheim in the academic year 1965–66 (Karbić, 2017, 343–345).

Most frequently, the journal published discussions on population, livestock, forestry, transport, and industrialization—that is, on the research questions that were also the focus of the conferences. Most of the authors examined only the historical phenomena and reviewed books printed in their own republic. Despite the journal's predominant focus on domestic scholarly production, the review section was important because it also included reviews of works by foreign authors, thus providing partial insight into the international arena and international trends in economic historiography. From this point of view, the reports on international congresses and conferences, in which Yugoslav economic historians were rarely able to participate, were also important. The international dimension of the journal was further enhanced by its opening to foreign colleagues. Michael Palairet, a specialist in Balkan history, and Halil Inalcik, historian of the Ottoman Empire, were among the most prominent contributors. In this way, the journal sought to promote the integration of economic historians into the international arena (Cvetković & Matković, 2021, 203–218).

Methodological Concepts and International Cooperation

Schematically, international cooperation took place in two ways: through the exchange of ideas and through participation in international conferences and congresses. A special type of international cooperation was cooperation with non-aligned countries. In international cooperation, the accessibility of international literature was crucial for following international trends and, consequently, for integration into international scientific communities. International literature was acquired through exchange or purchase. National and university libraries as well as academies of sciences played an important role in the exchange. Publications were exchanged with major Western universities and journals. Although purchases of Western literature were financed by the state, foreign aid also played an important role. For example, for several years the Rockefeller Foundation financed subscriptions to 40 American scientific journals, which were distributed to Yugoslav universities and institutes according to their own wishes. Scientific institutions also took advantage of this opportunity to acquire the volumes they could not obtain during World War II or because of the strong orientation toward the Soviet scientific press in the first postwar years (Bondžić, 2011, 61–138; Lloyd, 1970, 376–395).

The Definition of the Field

As in any other discipline, in the beginning of economic historiography there were questions about the definition of the field of research. In 1948 Nikola Vučo (1902–1993) published the first economic history of Yugoslavia. The book established the basic context. Vučo understood the history of Yugoslavia as the sum of the regional histories of the Yugoslav nations. Any economic history research had to begin at the level of individual nations (republics), and only then was a synthesis at the Yugoslav level possible. The basic framework of periodization and thus interpretation was an ideologically predetermined model of social development from prehistoric communities through feudalism and capitalism to socialism. Vučo saw the tasks of economic history in a clear Marxist manner. He focused on the study of modes of production and relations in historical perspective, the impact on social relations, class formation and class struggles, and the distribution of socially produced wealth and consumption. He also examined producers (workers and peasants) in light of technological development, giving priority to social structures (organizations, groups, classes, society) over the individual. In methodological terms, in keeping with the spirit of the times, he stressed the importance of historical (dialectical) materialism and the primacy of the material world and relations with other social systems. He also stressed that the use of both quantitative and qualitative methods was essential for any research, without going into these issues in detail (Vučo, 1948, 3–13).

Methodological issues came to the fore in a discussion between Rudolf Bićanić (1905–1968) of the University of Zagreb and the eminent Slovenian historian Fran Zwitter (1905–1988) of the University of Ljubljana. In 1952, Bićanić published an article on economic history at the interface between economics and history. He argued that the position of economic history between economics and historiography was contested. While he had no doubt that economic history was a historical science, he had many reservations as an economist. He based his observations on the then-current debates about economic history among economists (Fourastié, 1950; Nogarro, 1950; Sombart, 1902; Usher, 1951). He provocatively claimed that economic history was a branch of economics and not a historical science in the strict sense. Economic history was supposed to study how economic laws manifested themselves throughout history, and it supposedly leaned more toward abstraction, synthesis, and long-term observation of processes. Historians were supposed to prepare and organize the relevant historical material for further analytical consideration. Only the methodological tools of economics, he said, allowed real insights into the economic past. Econometric methods were to make economic history as a science the most accurate social science—within the limits of the Marxist conception, of course. Description and the search for causality—the hallmarks of historiography—were completely inadequate, Bićanić argued (Bićanić, 1952/1953, 824). Fran Zwitter, a historian who already sought a methodological synergy between historiography and sociology (Zwitter, 1938), strongly opposed the idea that economic history was not part of historical science. He also disagreed that historiography should be merely descriptive and stressed the importance of analysis and quantitative methods. He believed that the scholarship of economic history must be both grounded in theory and based on historical sources. The dual nature of economic history in terms of economists and historians seemed to him complementary and useful for the advancement of economic history as a discipline (Zwitter, 1954, 169–172).

In contrast to Bićanić, Mijo Mirković (1898–1963) took a broader view of economic history. He did not doubt that economic history was a historical science. Therefore, he also paid attention to historical sources. He promoted the publication of sources on the economic history of Croatia (Petrić, 2014, 101–107). Mirković saw economic history as a discipline that studies the economic past in its entirety and in an interactive relationship with society. He defined economic history as a "concrete" discipline because it encompasses a clear field of study: the state, the nation, the local community, and so forth, as indicated by the historical sources. He firmly believed that because of the profound differences in the social and economic development of the Yugoslav nations, economic history had to be studied in two ways. He joined Vučo in claiming that Yugoslav economic history is only the synthesis of the regional histories. This process was to reveal the "theoretical" character of economic history, since as a scientific discipline it had to generalize the results of research and search for the "laws of progress"—also by resorting to the methodological repertoire of economic sciences, especially in the field of quantification. Despite his expertise in economics, Mirković leaned toward traditional historical methods in his approach (Mirković, 1958, 1).

As a professor of economic history at the Faculty of Law, Rudolf Bićanić actively pursued research in economic history (Karaula, 2016, 216–242). Among the central research questions that preoccupied him were those concerning the emergence and development of capitalism. On this topic, he wrote two fundamental studies, one on Croatian manufactory production from the mid-eighteenth to the mid-nineteenth centuries and the other on the beginnings of capitalism in Croatian economics, economic thought, and politics (Petrić, 2014, 101–107). Bićanić turned to economic history when, due to disagreements with the authorities, he was unable to publish further works about contemporary Yugoslav economics, which was his main interest. After the publication of his book on the era of manufactories in Croatia, he was criticized that the book was not sufficiently based on Marxist theory. His idea to focus on the Industrial Revolution and the development of economic thought in Croatia remained unrealized. In the early 1950s, when intellectual life began to relax with the beginning of the liberalization of the communist regime, he was able to start researching modern economics. His papers were published in national and international journals. As a guest, he lectured at Western universities and international conferences. He spent two years in the United States at the University of Texas at Austin and at the Center for Advanced Studies in Behavioral Sciences in Palo Alto, California. His extensive social network enabled him to organize the publication of his book in the United Kingdom with Cambridge University Press. Professor Michael Kaser (1926–2021) (1973, VII–VIII), an economist and historian from St. Antony's College, Oxford, wrote the forward. Unfortunately, Bićanić died before completing the manuscript, and the book was edited for the press by his collaborators. The work, entitled *Economic Policy in Socialist Yugoslavia,* presents the history of post-World War II economic policy (Bićanić, 1973). It is strongly historical in nature; already the first chapter examines the economic and historical legacy and the context created by the establishment of the Yugoslav state after the First World War.

If economic history as a discipline was still poorly defined and underdeveloped in the late 1950s, the situation changed in the 1960s. Historians also began to pay more attention to economic history. In 1960, the subject of economic history was introduced at the Department of History at the University of Zagreb. It was taught by Igor Karaman (1927–1995), who had broad research interests. His contribution to the guidance of students at the postgraduate level was considerable. It is possible to generalize his research work to the tracing of long-term economic and social structural changes (Vranješ Šoljan, 1995, 7–9). Many years of research and a few smaller studies enabled him to summarize Croatian economic development from the beginning of the nineteenth century until World War II from the point of view of industrialization. The book also shows the absence of a Marxist approach and the replacement of that conceptual framework by modern historiographical concepts such as proto-industrialization (Karaman, 1991).

At the University of Belgrade, Nikola Vučo played an indispensable role in the process of establishing economic history as a discipline. As a researcher and professor of economic history at the Faculty of Economics (Aleksić, 2018, 214), he influenced many economic historians. His body of work is notable for combining national (that is, Serbian), Yugoslav, and international approaches in his research. His first book began with a synthesis of Yugoslav economic history (Vučo, 1948), while later he continued with the economic history of Serbia up to the First World War (Vučo, 1955). Both works were still embedded in the Marxist pattern of knowledge. In his later works, Vučo began to depart from this pattern. In doing so, he followed the ongoing process in historiography. Vučo also focused on general surveys of economic history. Toward the end of the 1950s, he published a book entitled *Ekonomska istorija* [*Economic History*]. The rather general title denoted a work on global economic history. In terms of periodization and problem structure, Vučo's work still conformed to the Marxist periodization of socioeconomic systems. He drew on a wide range of literature, with a slight preponderance of Marxist authors, especially of Soviet origin. However, he also referred to contemporary literature of the Western world—French, German, and English—which shows that he followed journals in different languages and thus with different methodological orientations (Vučo, 1959). A few years later, an updated edition of his book was published under the title *Ekonomska istorija sveta* [*Economic History of the World*] (1962). The book consisted of two parts, the most extensive of which dealt with the economic history of the Western world, that is, the development of the capitalist economy.

The second part was devoted to the economic history of Russia and finally to the Soviet Union. The second edition of the *Economic History of the World* deviated from the Marxist conception in its structure and conception. The periodization was no longer oriented to socioeconomic systems, but to the dynamics of development of the economy (Vučo, 1962). Both works were conceived in the spirit of Eurocentrism and adhered to the Western model of economic history. Vučo also addressed the relationship between the state and the economy. Except for the chapter on the Soviet Union, the book relied entirely on contemporary Western literature or sources published by Western institutions (Vučo, 1975).

In considering the development of economic history as a discipline, two economists who were originally from Yugoslavia but who worked in American academic institutions need to be mentioned. Their books provided a counterpoint to the emerging economic historiography, as they assumed an ambitious historical synthesis. Their works were available in Yugoslav libraries and were well-known to the professional public. However, it would be an exaggeration to claim that they had a great influence among economic historians. Jozo Tomasevich (1908–1994) was a native of Croatia. He was an economist and earned his doctorate in Basel, Switzerland. In the 1930s, between the two wars, he worked for the Yugoslav central bank. Just before World War II he was accepted into the Rockefeller Fellowship program and traveled to the United States, where he settled permanently and became the leading expert on the economic and political history of Yugoslavia (Vucinich, 1955, 257–258). In the field of economic history, his book ***Peasants, Politics, and Economic Change in Yugoslavia***, published by Stanford and Oxford University Press in 1955, is noteworthy. The first chapter presented the development of agriculture and the underlying economic and social problems of rural areas against the background of general economic changes. The second part dealt with the First World War, while the third part focused on the interwar period. The first part was conceived regionally, that is, the author analyzed the subject matter according to individual Yugoslav regions (republics), which he explained with very different pasts. In the other two parts, however, he changed the perspective and analyzed Yugoslavia as a whole, with a strong sense of regional specificity. Methodologically, the book followed the model that was later adopted in Yugoslav economic historiography. It was based on the traditional historical (positivist) model with emphasis on empirical evidence, insightful descriptive analysis, and partial incorporation of quantitative methods as far as the source material allowed (Tomasevich, 1955).

The second author, Toussaint Hočevar (1927–1987), was originally from Slovenia. After World War II he emigrated to Austria, where he studied economics in Innsbruck and obtained his PhD before settling in the United States as a professor of economics at various universities, finally at the University of New Orleans (Granda, 1988, 129–131). He entered Slovenian economic historiography with a survey of Slovenia's economic history from the abolition of serfdom to the early 1960s. As an economist, Hočevar based his work on economic analysis in a historical context. As early as 1965, he described in detail the consequences of Slovenia's accession to the Yugoslav state and the change in Slovenia's relative economic position within the Yugoslav state up to the 1960s (Hočevar, 1964, 1965). Hočevar went one step further and wrote his next book, which was published in the United States, in Slovenian. In the work, titled *Slovenski družbeni razvoj* [*Slovenian Social Development*], he posed the question of the connection among the economic structure, development dynamics, and the ethnic emancipation of the Slovenes in the nineteenth century. He addressed the question of the importance of language for the transfer of technology and knowledge and examined the "economic costs" of Slovenian bilingualism (Slovene, German) (Hočevar, 1979). Hočevar was far ahead of Slovenian historiography, on both the conceptual and the methodological level (Lazarević, 2012, 116–136).

Ivan Erceg, the editor of the journal *Acta Historico-Oeconomica Iugoslaviae*, felt increasingly uncomfortable with the general development of economic history as a discipline. For him, it was clear that Yugoslav economic historiography had problems originating from being trapped in its own (republican) frame of reference. As a result, international research and methodological trends were not systematically monitored or implemented. He missed broader international cooperation, frequent participation in international conferences, and more intense engagement in international organizations (Cvetković & Matković, 2021, 203–218). By the 1980s economic historiography in Yugoslavia had consolidated institutionally and in terms of personnel. The discipline faced the challenge of a qualitative transformation—to update its conceptual and methodological foundations. Erceg called for renewal in a situation in which the communist regime was losing its legitimacy due to a profound economic crisis, accompanied by the disintegration of Yugoslav intellectual communities (Stallaerts, 2002, 141–157).

Erceg's disappointment stemmed from the fact that (economic) historians had every opportunity to learn about international historiographical trends. Cultural transfer was possible, as evidenced by popular culture (Janjetović, 2011; Vučetić, 2012). Since the beginning of the 1960s, Yugoslavia permitted free travel, and there were large migrations to the Western countries. The import of foreign literature, historical journals, and books was allowed, and these publications were available to a wide range of readers in academic and national libraries. Books and periodicals could also be ordered from local booksellers. Historians traveled abroad and visited libraries and bookstores in Western Europe. This is also evidenced by the history journals, which regularly published news or reviews of new books and reports on international conferences and congresses. They bear witness to the fact that some historians regularly followed contemporary historical production and disseminated their findings to the wider community of their professional colleagues (Aranitović, 1999; Najbar Agičič, 2013, 274–300).

International Cooperation

Given liberalization, it is surprising that hardly any historians studied abroad (Perišić, 2008, 396–419), compared to economists (Bockman, 2011, 77–106) or sociologists (Naumann, 2018, 97–122). Before World War II, the Yugoslav intelligentsia was predominantly educated abroad.

On the eve of the war Yugoslav students studied in France (146), Germany (including Austria as part of the Reich, 312), Czechoslovakia (193), Italy (98), and Great Britain (36). France, Austria, and Germany were crucial for the development of Yugoslav science (Bondžić, 2011, 46–47). The heavy reliance on the Soviet Union in the early postwar years marked a departure from the previous practice of cooperation with several universities and institutes throughout Europe. International cooperation was in the hands of the federal government ministry responsible for science. Invitations to congresses or other international events as well as offers of cooperation were arranged through the diplomatic missions abroad. The cultural attachés were also responsible for the field of science. Their task was to establish contacts with scientific institutions and to look for opportunities for cooperation. They were also responsible for guest appearances by scientists in other countries. All visits of Yugoslav scientists (and students) abroad and of foreign scientists in Yugoslavia had to be organized and registered. The same applied to student exchanges.

The reorientation of scientific policy also affected the possibilities for international cooperation. Immediately after World War II, 86 percent of Yugoslav students abroad were in Eastern Bloc countries, mainly in the Soviet Union or Czechoslovakia. The picture of Soviet domination becomes even clearer if one considers the number of military scholarship students in the Soviet Union, as they exceeded the number of civilian students. Yugoslav students participated in all fields of study, but up to 60 percent of them concentrated in technical fields (mining, transport, agriculture, forestry). Among the other subjects, the study of economics and medicine stood out. This was also a clear expression of the priorities of Yugoslav development policy in the postwar years. The Soviet Union provided Yugoslavia with most of the aid, professors, and other personnel needed to organize the study programs and scientific activities in the country.

Although cooperation with the Soviet Union covered most of the activities, cooperation with Western countries did not cease completely. Students continued to attend Western universities, although in much smaller numbers than before (Perišić, 2008, 83–92, 208). However, the traditions and established connections could not be eliminated. An example of this was the visit of the eminent British historian A. J. P. Taylor to Yugoslav universities in 1947 (Bondžić, 2011, 61–80). Yugoslav scientists regularly attended various congresses, and individuals were sent to Western countries for training and study purposes. If the participation of scientists from the Soviet Union or other communist countries was guaranteed, Yugoslavia had no objection to Yugoslav scientists attending as well. However, any visit abroad had to be announced and approved by the authorities and invitations to congresses accepted through diplomatic missions.

After 1948 everything changed due to political conflict with the Soviet Union. Scientific and other forms of cooperation with socialist countries ceased. There were no more exchanges or cooperation on scientific projects. Science was in the same situation as the whole country. The way out of isolation was to the West. Yugoslavia received economic, military, and political aid from Western countries during the conflict with the Soviet Union (Lees, 1997). Gradually, scientific cooperation with Western countries began to intensify. It depended not only on the will of Yugoslavia but also on the willingness of Western countries to promote cooperation. Once, the initial mistrust was overcome and political relations were settled, cooperation with Western institutions became smoother. Opportunities for scientific cooperation arose through the national institutions of the West (British Council; Rockefeller, Carnegie, and Ford Foundations; grants from Western governments) and international organizations (UN, UNESCO, OECD, various international associations, academies of sciences and arts). Authorities now urged the scientific community to cooperate as much as possible with Western institutions in the name of scientific progress and faster development of the country. Scientists traveled to international conferences in an organized manner, as a kind of national delegation. Care was taken to send scientists abroad who spoke foreign languages and had comparable scientific achievements. The norm was to be fluent in at least one world language to communicate with the international community. The scope of cooperation grew enormously and did not go only in one direction: foreign scientists visited Yugoslav universities and institutes much more frequently than before, and the country offered foreign students many opportunities to live and study there.

By promoting international cooperation, Yugoslavia sought to accelerate the transfer of knowledge and technology. There was also a desire to expand the international social network of the scientific community. Individual scholars began to travel abroad independently or at the expense of their institutions in 1950s. One of the main obstacles to collaboration was the lack of resources, especially for longer study trips and specializations. Technical assistance from Western governments, which offered several study grants or funded short stays for students, contributed significantly to the integration of Yugoslav science into the international scientific landscape (Bondžić, 2011, 61–138; Lloyd, 1970, 376–395).

The number of Yugoslav scientists traveling abroad increased and became more geographically dispersed. In 1957, 27.5 percent of Yugoslav university teachers traveled abroad (to the West) for congresses, short specialization courses, or study tours (Bondžić, 2018, 213). The level of cooperation continued the tendencies that were traditionally present in the education of the Yugoslav scientific community even before World War II (Perišić, 2008). Universities from Central Europe (West Germany, Austria, Switzerland, Italy), supplemented by French and British universities, maintained their pre-eminence in the postwar period (Bondžić, 2011, 154). Later—in the 1960s—the United States also gained importance. By the 1970s the United States' share was already dominant (Naučna politika, 1980, 281). The role of the United States grew thanks to the various foundations for scientific and cultural cooperation. Particularly noteworthy were the Ford Foundation (from 1959) and the Fulbright scholar and professor program (Barghoorn, 1967, 116). Both programs involved scholars who passed on their acquired knowledge to the Yugoslav society and economy. A study of the economists who studied in the United States, especially through Ford Foundation, has shown that after their return their teaching influence (neoclassical economics) was considerable—both on the formation of development policy and on the content of university studies of economics (Bockman, 2011, 77–106; Borak, 2019). It is reasonable to expect that the situation was similar in other disciplines.

In the decade between 1953 and 1963, some 17,000 experts traveled abroad for training (doctorates, specializations) and study tours (visits to institutions, conferences, congresses) (Bondžić, 2011, 159). By 1978, about 30,000 scientists, students, professionals in training, and technical experts had participated in international scientific exchange programs in the United States and Western Europe (Naučna politika, 1980, 281). This figure does not include those who traveled on their own for study purposes without informing the authorities of their departure (Perišić, 2008, 396–419).

In 1980, Yugoslav scientific institutions cooperated with institutions from 20 countries: 110 projects were carried out with Western European and other developed countries, and 114 projects were realized in cooperation with institutions from the United States. Yugoslavia also actively cooperated with the OECD and European Economic Community. On the one hand, it was a recipient of technical assistance, and on the other, Yugoslav scientists participated in joint or international projects. In 1980, they were involved in 22 national projects and eight joint projects in the framework of the OECD. In the framework of scientific cooperation with the European Economic Community, Yugoslavia was included in the COST program from 1971. In 1980, it participated in six international scientific projects, and in the same year Yugoslav scientific institutions joined another twelve COST international scientific projects.

After relations with the Soviet Union were restored in the second half of the 1950s, cooperation with East European communist countries was resumed. However, its scope did not reach the level of cooperation with Western scientific and other institutions. The reorientation toward Western science was permanent. While cooperation with Western institutions increasingly fell within the sphere of interests of individual Yugoslav universities, institutes and even individuals, cooperation with communist countries of Eastern Europe, especially the Soviet Union, was still highly regulated (Bondžić, 2011, 61–138; Lloyd, 1970, 376–390). In the eastern part of Europe, Yugoslavia was able to pursue a soft power policy because of its proximity to the Western world (Vuletic, 2012, 115–131). From 1964, when an agreement on scientific and technical cooperation was signed, until 1977, Yugoslav institutions participated in a total of 250 projects with institutions from East European countries—which means that it took thirteen years to reach the volume of cooperation with Western countries in the single year 1980. The exchange of researchers became more frequent, but its frequency was far below the cooperation with the Western institutions. In 1977 only 400 people traveled in both directions (Naučna politika, 1980, 269–285).

Cooperation with Non-Aligned Movement Countries

Yugoslav international policy was based on the principles of non-alignment, and Yugoslavia was one of the founding and leading countries of Non-Aligned Movement (hereafter the NAM) (Mišković et al., 2014). This movement also affected scientific cooperation. In the mid-1950s increased cooperation with the countries of the so-called Third World began. Yugoslavia's role in the NAM opened opportunities for international scientific exchange. In this context, Yugoslavia found itself in a different position than in its relations with the West, where it was the recipient of international technical assistance: it could provide expert help in the transfer of knowledge and technology to Third World countries. As part of the reorientation toward the West, a systematic scholarship policy for foreign students was introduced. In the early 1950s, students initially came from West European countries, usually to study Yugoslav languages, culture, literature, art, and history. This lasted until the dissolution of the Yugoslav state.

Gradually, opportunities arose for students from Asia and Africa to study in Yugoslavia. Students from these countries were also expected to promote economic cooperation with Asian and African countries (Bondžić, 2014, 637–648). By 1980 nearly 7,000 of them had received Yugoslav scholarships, and 2,600 were studying at Yugoslav universities at their own expense. More than 1,000 participated in various short-term training and specialization programs. This was not the final figure, however, as many experts remained in Yugoslavia and continued their training in various enterprises under economic cooperation or technology transfer programs. Between 1954 and 1980 Yugoslavia sent over 9,000 experts to developing countries. They provided technical assistance in the areas of greatest need for the countries; many of which were just gaining statehood and needed support in organizing their societies and economies in the post-colonial era (Naučna politika, 1980, 281). The cooperation with NAM countries was based on technical and economic field; there was simply not enough room nor broader interest for historians or economic historians to be included.

The economic model of self-government attracted the attention of the international intellectual community (Bockman, 2011, 77–80; Woodward, 1995, 191–221). The idea that the Yugoslav model could be exported to developing countries was also gaining ground among Yugoslav policymakers. This led to the establishment of the International Center for Public Enterprises in Developing Countries in Ljubljana. This was the only such institution within the NAM, established in 1974 by the Yugoslav government in cooperation with the United Nations. More than 40 countries joined and participated in the work of this center, which provided expert assistance, training, research, information support and experts who could advise developing country governments. This happened at a time when Yugoslavia enjoyed a high reputation, and its model of market socialism was accepted as an alternative to the Soviet planned economy. The center promoted the training of state enterprise executives, global financial cooperation among developing countries, industrial restructuring, the division of large monopoly enterprises into smaller enterprises, and the creation of small- and medium-sized enterprises. It also promoted entrepreneurship to manage publicly owned enterprises (Bockman, 2011, 102–104). The centers in other parts of the country that were involved in agriculture and fisheries must also be considered. The Agribusiness Development Center, the Maize Center, and the Fishery Center also worked with Third World countries. Meanwhile, a systematic study of developing countries was carried out by the Institute of Developing Countries (Naučna politika, 1980, 281; Polič & Sajko, 1973, 97–101).

Conclusion

Economic historians were on the margins of the historiographical community and public interest. The restriction to regional (republican) contexts or the isolated study of phenomena of local significance within a narrow temporal framework led to a certain "self-sufficiency." These circumstances constituted an obstacle to the methodological and conceptual modernization of economic historiography, despite the opportunities for travel and liberalized cultural transfer. Only a few economic historians were involved in international cooperation. They participated in international historical commissions with Western and Eastern countries, accepted invitations to conferences, and traveled to economic historians' congresses. Bilateral historical commissions were established with Italy, Austria, France, Hungary, Czechoslovakia, the German Democratic Republic, and the Soviet Union. However, bilateral commissions with Spain, Greece, Poland, Bulgaria, and the Federal Republic of Germany failed to become fully operational. Yugoslav economic historians also participated in regional initiatives such as the Internationales Kulturhistorisches Symposion Mogersdorf as of 1969, which brought together historians from the border regions of Slovenia, Croatian, Austria, and Hungary (Mihelič, 2003, 577–587). The possibilities of international cooperation were not fully exploited—not only because of the methodological lag as stated by Ivan Erceg, but also because of latent financial problems. For this reason, economic historiography was understaffed and under-researched, which was one of the major difficulties in its involvement in the international arena.

Regarding international cooperation, it is important to mention that the Commission for Economic History was established with the desire to cooperate with the IEHA. Membership enabled Yugoslav economic historians to participate in the World Congresses of Economic History. Given the national organization of the IEHA at the time, this was a necessity (Berg, 2015, 56). In 1965 the Yugoslav Commission for Economic History became a member of the IEHA. In the same year a small group of Yugoslav economic historians participated in the WEHC in Munich. Later they also participated in the congresses in Bloomington (1968) and Leningrad (1970). Participation in the congresses was associated with considerable financial problems, as it was difficult to raise the necessary funds. Since the economic history organization had no financial resources of its own, it had to rely on the support of the republican and federal departments of international cooperation. In the 1970s the situation became critical. In 1978, Yugoslav experts finally managed to attend a congress in Edinburgh. Yugoslav economic historians even proposed to hold a congress in Dubrovnik in 1982 but withdrew their offer because they were unable to secure the necessary political and financial support in the country. In the 1980s they participated in the congresses in Budapest (1982) and Bern (1986) (Cvetković & Matković, 2021, 212–218).

References

Aleksić, V. (2018). Ekonomska istorija Srbije u domačoj istoriografiji 2010–2017: metodološki izzazovi u kontekstu interisciplinarnih istraživanja. *Ekonomska i Ekohistorija, 14*, 212–224.

Aranitović, D. (1999). *Jugoslovenski istorijski časopis - Bibliografija 1935 – 1997*. Beograd.

Banac, I. (1992). Historiography of Countries of Eastern Europe: Yugoslavia. *The American Historical Review, 97*(4), 1084–1104.

Banac, I. (1995). The Dissolution of Yugoslav Historiography. In S. Ramet & L. Adamovich (Eds.), *Beyond Yugoslavia: Politics, Economics, and Culture in a Shattered Community* (pp. 39–65). Westview Press.

Barghoorn, F. (1967). Cultural Exchange Between the Communist Countries and the United States. *The Annals of the American Academy of Political and Social Science, 372*, 113–123.

Berg, M. (2015). East-West Dialogues: Economic Historians, the Cold War, and Detente. *The Journal of Modern History, 87*(1), 36–71.

Bićanić, R. (1952/53). Ekonomska povijest medju ekonomskim in historijskim naukama. *Zgodovinski časopis, 6–7*, 820–825.

Bićanić, R. (1973). *Economic Policy in Socialist Yugoslavia*. Cambridge University Press.

Bockman, J. (2011). *Markets in the Name of Socialism The Left-Wing Origins of Neoliberalism*. Standford University Press.
Bondžić, D. (2011). *Misao bez pasoša - Međunarodna saradnja beogradskog univerziteta 1945–1960*. ISI.
Bondžić, D. (2014). Školovanje studenanta iz zemalja u razvoju kako deo spoljne politike Jugoslavije 1950–1961. *Annales, 24*(4), 637–648.
Bondžić, D. (2018). Razvoj nauke u Jugoslaviji posle drugog svetskog rata. *Kultura, 161*, 201–220.
Borak, N. (2019). *Študij ekonomije v Jugoslaviji*. Zveza ekonomistov Slovenije.
Brunnbauer, U. (2003). Kontinuitete in spremembe. Aktualni trendi v zgodovinopisju Jugovzhodne Evrope. *Zgodovinski časopis, 57*(3–4), 431–450.
Cvetković, E., & Matković, A. (2021). Rađanje jedne istoriografije: nasleđe Komisije za ekonomsku istoriju Jugoslavije. In A. V. Matković & A. Miljković (Eds.), *Izazovi izučavanja ekonomske istorije u Srbiji* (pp. 197–223). Institut ekonomskih nauka.
Dragišić, P. (2019). *Zidari tuđe sreće. Ekonomska emigracija iz Jugoslavije u Zapadnu Evropu krajem šestdesetih i početkom sedamdesetih godina*. INIS.
Fourastié, J. (1950). *Histoire économique*. IXe congrès international des sciences historiques.
Gabrič, A. (2008). Znanstvena politika v Sloveniji po drugi svetovni vojni in vloga Antona Peterlina. In *Anton Peterlin 1908–1993* (pp. 300–330). Slovenska akademija znanosti in umetnosti, Inštitut Jožef Stefan.
Gabrič, A. (2009). Med idejami proučevanja novejše zgodovine in zgodovine delavskega gibanja. In A. Gabrič (Ed.), *Zgodovinopisje v zrcalu - 50 let Inštituta za novejšo zgodovino* (pp. 17–48). INZ.
Gabrič, A. (Ed.). (2010). *Prežihov Voranc - Lovro Kuhar: pisatelj, politik, patriot*. INZ.
Grafenauer, B. (1947). Problemi in naloge slovenskega zgodovinopisja v našem času. *Zgodovinski Časopis, 1*(1–4), 11–28.
Granda, S. (1988). Toussaint Hočevar (25.6.1927–21.4.1987). *Zgodovinski časopis, 42*(1), 129–131.
Hočevar, T. (1964). *Slovenian Role in Yugoslav Economy*. Slovenian Research Center.
Hočevar, T. (1965). *The Structure of Slovenian Economy*. Studia Slovenica.
Hočevar, T. (1979). *Slovenski družbeni razvoj*. Prometej.
Janjetović, Z. (2011). *Od internacionale do komerciale. Popularna kultura u Jugoslaviji 1045–1991*. INIS.
Karaman, I. (1991). *Industrializacija građanske Hrvatske 1800 – 1941*. Naprijed.
Karaula, Ž. (2016). Prilozi za biografiju Rudolfa Bićanića. In *Radovi zavoda za znanstveno istraživački i umetnički rad u Bjelovaru, 10*, 216–242.
Karbić, D. (2017). Ivan Erceg (1919–2017). *Zbornik Odsjeka Povijesnih Znanosti HAZU, 35*, 343–345.

Kaser, M. (1973). Foreword. In R. Bićanić (Ed.), *Economic Policy in Socialist Yugoslavia* (pp. VII–VIII). Cambridge University Press.

Kresal, F. (1973). Znanstveno posvetovanje o industrijski revoluciji v jugoslovanskih deželah. *Prispevki Za Zgodovino Delavskega Gibanja, 13*(1–2), 313–317.

Lazarević, Ž. (2012). Identitete in imena gospodarskega zgodovinopisja v Sloveniji. *Ekonomska i Eko Historija, 8*, 116–136.

Lees, L. (1997). *Keeping Tito Afloat: The United States, Yugoslavia and the Cold War.* The Pennsylvania State University Press.

Lloyd, J. (1970). Scientific and Technological Cooperation among East European Communist Countries. *Minerva, 8*(3), 376–395.

Lukšič, H. M. (2020). Beg možganov od konca petdesetih do začetka devetdesetih let 20. stoletja s poudarkom na Sloveniji. *Dve Domovini/two Homelands, 52*, 113–128.

Malsimović, D. (1968). Yugoslavia: Development of Scientific Research. *Nature, 218*, 846–849.

Marković, P. (2004). Kako smo pronašli "pravu istorijsku istinu"? Srpska istoriografija posle 1991. godine. *Prispevki za novejšo zgodovino, 44*(2), 45–66.

Mihelič, D. (2003). Od sobivanja do razhoda. Utrinki iz zadnjih dveh desetletij sodelovanja zgodovinarjev druge Jugoslavije. In L. O. Perovšek (Ed.), *Zbornik Janka Pleterskega* (pp. 573–595). Znanstveno raziskovalni center.

Mirković, M. (1958). *Ekonomska historija Jugoslavije.* Ekonomski pregled.

Mišković, N., Fischer-Tiné, H., & Boškovska, N. (Eds.). (2014). *The Non-Aligned Movement and the Cold War* (Delhi – Bandung – Belgrade). Routledge.

Najbar, A. (2013). *U skladu s marksizmom ili činjenicama. Hrvatska historiografija 1945–1960.* Ibis.

Naučna politika i naučno-istraživačka delatnost. (1980). *Jugoslovenski pregled, 24*(7–8), 269–285.

Naumann, K. (2018). International Research Planning Across the Iron Curtain: East-Central European Social Scientiests in the ISSC an Vienna Centre. In M. Christian, S. Kott, & O. Matejka (Eds.), *Planning in Cold War Europe: Competition, Cooperation, Circulations (1950s–1970s).* Walter De Gruyter.

Nogaro, B. (1950). *La méthode de l'économie politique.* Librairie Générale de Droit et de Jurisprudence.

Perišić, M. (2008). *Od Staljina ka Sartru. Formiranje jugoslovenske inteligencije na evropskim univerzitetima 1945–1958.* INIS.

Petrić, H. (2014). Prilozi poznavanju historiografije o gospodarskoj povijesti u Hrvatskoj. In M. Darja (Ed.), *Vizija raziskav slovenske gospodarske in družbene zgodovine* (pp. 97–124). Založba ZRC.

Polič, M., & Sajko, K. (1973). Africa Research Institution in Yugoslavia. *Africa Spectrum, 6*(1), 97–101.

Prežihov, V., (Kuhar, L.). (1945). *Jamnica: roman soseske*. Slovenski knjižni zavod.

Ramnath, N. (1988). Yugoslavia: Self-management or Mismanagement? *Economic and Political Weekly, 23*(40), 2052–2054.

Repe, B. (1996). Razpad historiografije, ki nikoli ni obstajala. *Zgodovina Za Vse, 1*(3), 69–78.

Repe, B. (1999). Jugoslovanska historiografija po drugi svetovni vojni. *Tokovi Istorije, 1–4*, 312–325.

Režek, M. (2014). Usmerjena preteklost: Mehanizmi ideološke in politične "kontaminacije" zgodovinopisja v socialistični Sloveniji in Jugoslaviji (1945–1966). *Acta Histriae, 22*(4), 971–992.

Šoljan, N. N. (1991). The Saga of Higher Education in Yugoslavia: Beyond the Myths of a Self-Management Socialist Society. *Comparative Education Review, 35*(1), 131–153.

Sombart, W. (1902). *Der Moderne Kapitalismus: Historisch-systematische Darstellung des gesamteuropäischen Wirtschaftsleben von seinen Anfängen bis zur Gegenwart*. Duncker & Humblot.

Stallaerts, R. (1999). Historiography in the Former and New Yugoslavia. *Belgisch Tijdschrift Voor Nieuwste Geschiedenis, 3–4*, 315–336.

Stallaerts, R. (2002). The Disintegration of the Yugoslav Intellectual Community In B. Coppieters & M. Huysseune (Eds.), *Secession, History, and the Social Sciences* (pp. 141–157). VUB University Press.

Tomasevich, J. (1955). *Peasant, Politics and Economic Change in Yugoslavia*. Stanford University Press and Oxford University Press.

Usher, A. P. (1951). Sir John Howard Clapham and the Empirical Reaction in Economic History. *The Journal of Economic History, 11*(2), 148–153.

Voje, I. (2004). Prof. Dr. Jadran Ferluga (1920–2004). *Zgodovinski časopis, 58*(1–2), 207–209.

Vranješ Šoljan, B. (1995). Prof. dr. Igor Karaman (1927–1995). *Radovi, 28*(1), 7–9.

Vučetić, R. (2012). *Koka-kola socializam. Amerikanizacija jugoslovenske popularne kulture šestdesetih godina XX veka*. Službeni glasnik.

Vucinich, A. (1955). Jozo Tomasevich: 1908–1994. *Slavic Review, 54*(1), 257–258.

Vučo, N. (1948). *Privredna istorija naroda FNRJ do prvog svetskog rata*. Naučna knjiga.

Vučo, N. (1955). *Privredna istorija Srbije do prvog svetskog rata*. Naučna knjiga.

Vučo, N. (1959). *Ekonomska istorija*. Naučna knjiga.

Vučo, N. (1962). *Ekonomska istorija sveta*. Naučna knjiga.

Vučo, N. (1975). *Državna intervencija u privredi*. Savremena administracija.

Vuletic, D. (2012). Sounds Like Amerika. Yugoslavias Soft Power in Eastern Europe In P. Romijn, G. Scott-Smith, & J. Segal (Eds.), *Divided Dreamworlds: The Cultural Cold War in East and West* (pp. 115–131). Amsterdam University Press.

Woodward, S. (1995). *Socialist Unemployment. The Political Economy of Yugoslavia 1945–1990*. Princeton University Press.

Zwitter, F. (1938). *Sociologija in zgodovina*. Akademska založba.

Zwitter, F. (1954). Gospodarska zgodovina kot znanost. *Zgodovinski časopis, 8*(1–4), 169–172.

CHAPTER 8

Romania: Approaches and Limits of Economic History

Bogdan Murgescu, Matei Gheboianu, and Ionuț-Mircea Marcu

In this chapter, we will outline the way Romanian historians who approached economic themes dealt with the constraints of the institutional framework and tried to overcome the limitations imposed by the communist regime on international cooperation in the field of social sciences and humanities. We will investigate the (few) connections across the Iron Curtain, the attempts to connect Romanian historiography to its Western counterparts, and the limits of the partial opening which started in the late 1950s and continued during the 1960s and a part

B. Murgescu (✉) · M. Gheboianu · I.-M. Marcu
University of Bucharest, Bucharest, Romania
e-mail: bogdan.murgescu@istorie.unibuc.ro

M. Gheboianu
e-mail: matei.gheboianu@istorie.unibuc.ro

I.-M. Marcu
e-mail: ionut.mircea.marcu@drd.unibuc.ro

A. Dolezalova and C. Albrecht (eds.), *Behind the Iron Curtain*, Palgrave Studies in Economic History,
https://doi.org/10.1007/978-3-031-31578-7_8

of the 1970s. In this framework, we will focus on the case of Andrei Oțetea (1894–1977), but also describe the experiences of other scholars like Gheorghe Zane (1897–1978), and we will try to assess what these personal experiences reveal about the more general field of economic history.

As in many other countries, in interwar Romania historical and economic research were two different fields, with few entanglements. Consequently, economic history was not an integrated or autonomous academic field. While economists were rarely interested in economic history and at best considered only the recent past of the economic themes they discussed, research on economic history was conducted mainly by historians who came across interesting sources related to the economy of past times. Generally, these historians had no special training in economics, were not interested in methodological issues, and only seldom discussed the theoretical implications of their research. In the 1930s, several Romanian historians, such as Petre Panaitescu (1900–1967) or Andrei Oțetea, started to argue that some major historical processes were determined or at least heavily influenced by economic factors, and linked this conclusion either to Marxism or to other narratives of economic materialism (e.g., the works of Werner Sombart), but their approach remained marginal in the broader academia and focused mostly on domestic disputes, without significant links to the international developments in economic history (e.g., there was no Romanian scholar involved in the workings of the International Scientific Committee on Price History).

While the framework of Marxism imposed an economically centered approach on history, Romanian historians did not produce, with few exceptions, significant contributions on the economic history of the country. It can be argued that, consequently, for most Romanian historians economic history became intrinsically linked to this rudimentarily imposed Marxist *vulgata* (Murgescu, 2000, 34–35).

The general evolution of history writing was deeply embedded in the broader institutional context of social sciences and humanities in communist Romania. While there was no specific framework of economic historical research in communist Romania, the broader structural evolutions did have an impact on this somewhat underdeveloped field of research. Historical research was part of a centralized system of resource allocation, institutional and personal competition being very often the explanation of historians' positions in relation to the political field. On the other hand, there was a significant difference between the first part of the history of the Romanian communist regime, in which political interventions in historical debates and knowledge production were much more direct, and the last two decades, in which nationalism was used as the hegemonic master narrative by coopting large parts of the Romanian historiographical field.

While economic history was not one of the main fields of interest for Romanian historians, there were still several relevant debates in the context of a new official history treatise, with significant political and ideological importance, in which economic history arguments were employed or, in theory, could have been employed. Given that Marxism states the primordiality of the base over the superstructure, the new master narrative(s) imposed by Mihail Roller (1908–1958) had, at least in theory, an economic base. But given the insignificant training in economic methods and theory of most Romanian historians, as well as their preference for political history and positivism, the primordiality of economic structures in history remained only a *mantra* without any real epistemic progress, empirical or otherwise. While, in theory, economic history should have provided strong arguments for the participants in these debates, in reality, apart from superficially understood and applied Marxist concepts, economic structures were rarely used as means for proving or disproving an interpretation. In the 1970s and 1980s, with Marxism replaced by nationalism as the hegemonic analytical narrative in historical writing, the role of economic aspects became even less important. While political history became crucial to the regime's propaganda efforts, both within Romania, as well as abroad (Velimirovici, 2015a, 196–268), there was almost no ideological incentive to promote economic analyses.

Historians who undertook research on economic history and economists interested in economic history and in the history of economic thought communicated minimally. This relative segregation was determined in part by different chronological focuses—historians were more interested in the late medieval and early modern economic history, while economists rarely dealt with pre-1800 topics—and in part by the mutual prioritization of networks inside each of the two professions. In 1970 under the leadership of the Academy of Social and Political Sciences a Committee of Economic History and History of Economic Thought was established (mainly by economists). The committee became affiliated at the International Economic History Association (hereafter the IEHA) but did not succeed in bridging the professional divide between economists and historians or in setting up meaningful cooperative projects. Quite revealingly, the *Encyclopaedia of Romanian Historiography* published in 1978, which has a chapter dedicated to institutions with more than 100 entries, does not include any entry on the Committee of Economic History and History of Economic Thought or on the economic history chair of the Academy of Economic Studies (Enciclopedia istoriografiei…, 1978, 355–391).

The situation did not improve in late communism, when the Ceaușescu regime turned more nationalist and opportunities for academic cooperation were even more limited than before. In the historical profession interest in economic history declined, while economists focused even more on the practicalities of contemporary issues and produced only (some) contributions on the history of economic thought.

Phases in the Evolution of Romanian Historiography

After the end of World War II, the historiographical field went through structural changes. The post-1945 evolution of history writing is divided in the literature into three stages. The first one, from 1945 to the mid-1950s, was characterized by a significant control of the Romanian Communist Party, from 1948 to 1965 officially named Romanian Workers Party (hereafter the Party). The "liberal" phase, until the early 1970s, increased the number of professional opportunities available to Romanian historians, as several of them were coopted by the communist regime. The final stage, until the fall of the regime in 1989, signified a return to strict political control and decreasing agency for professional historians (Constantiniu, 2007; Iacob, 2011; Mihalache, 2003; Stan, 2010; Velimirovici, 2015a; Verdery, 1991; Zavatti, 2016).

The new historiographical master narrative imposed by Stalinist hardliner Mihail Roller during the first stage of the communist regime had almost no connection to the pre-1945 Romanian history writing, as it was based around the supposedly strong relations between Romanians and Slavs. In theory, this new hegemonic historiographical discourse was based on Marxism and Marxism-Leninism, but a short overview on the actual historical outputs disproves this idea. Actually, Romanian historians used only superficially some Marxist concepts, sometimes trying to fit the actual historical phenomena in a pre-determined Marxist historiographical *vulgata*. There are several explanations for the absence of real Marxist historiography in postwar Romania.

First of all, there was almost no tradition in ideological or theoretical debates in pre-communist Romanian history writing. Romanian historians were rarely involved in debates regarding how history ought to be defined methodologically and theoretically. Second, with several exceptions, such as Andrei Oțetea or Petre Panaitescu, Marxism as a theoretical framework of analysis was absent from Romanian history writing before World War II (Murgescu, 2000, 34–35). In economic history as a particular (sub)field, institutional changes were significant. As Marxism proclaimed that the economic basis determines the general evolution of society, historians had to refer to the economy in almost all studies they undertook. Yet, this enhanced interest in the economy was determined mainly by ideological conformism and was not backed by significant training on methodological issues or by a deeper understanding of economics; under these circumstances, inside history faculties in higher education institutions and inside history research institutes there was no institutional separation of units of economic history.

Second, the role of economics in the general framework of Romanian academia expanded. In Bucharest, the Academy of High Commercial and Industrial Studies founded 1913 was reorganized in 1948 into a full-fledged university (Institute for Economic Sciences and Planning, renamed in 1967 Academy of Economic Studies) with massive numbers of students and staff; at the same time, faculties of economics were established or expanded also in other major universities (Cluj, Iași, Timișoara, and so on). A centralized Institute of Economic Research [Institutul de Cercetări Economice] was founded in 1953, followed by several specialized institutes subordinated to various ministries and other administrative bodies (Zaman et al., 2013). Inside Academy of Economic Studies, the various universities teaching economics, and the Institute of Economic Research, economic history of the nineteenth-twentieth centuries was part of the curriculum and of the research structure, sometimes backed by distinct chairs and research units, but its role was rather marginal compared with present-oriented teaching and research. Starting with the late 1950s there was also an enhanced interest in the history of economic thought, which benefitted symbolically from the ideological interest to stress the links between communist nationalist economic policies and prestigious pre-communist economists who had argued for industrialization, for protectionism, or for other forms of sustained national economic development.

The entire institutional framework of history writing was reorganized after 1945, after the interwar Soviet model (Catalan, 2009). Research institutes, previously coordinated by universities, were transferred to the supervision of the Romanian Academy (officially renamed Academy of the People's Republic of Romania). Important historians, such as Gheorghe Brătianu (1898–1953) or Petre Panaitescu, suffered the consequences of their political involvement before and during the war, while the universities, research institutes, and the Academy were reformed. All those deemed as not sufficiently attached to the new regime were eliminated from research and academia. Other historians, such as Silviu Dragomir (1888–1962), Ion Nistor (1876–1962), Ioan Lupaş (1880–1967), Constantin C. Giurescu (1901–1977), lost (some temporarily) their professional positions (Opriş, 2006, 7). The new historiographical field was strictly controlled by Mihail Roller, who actively imposed a new historiographical master narrative and promoted his allies in research and academia (Bosomitu & Burcea, 2012).

The situation did not dramatically change with the fall of Roller at the end of the 1950s and the end of the Soviet-dependent period in Romanian historiography. While the fall of Roller is strongly linked both to events occurring within the high levels of power as well as to the tensions his rule created within the historiographical field (Velimirovici, 2015a, 48–75), it was also the start of a period defined as "liberal." Most of the formerly imprisoned or removed historians (those who survived, at least) were reinstated in various academic positions (Stoica, 2012). It also meant that a new group of historians could assert themselves in powerful institutional and symbolic positions, Andrei Oţetea being one of them (Iacob, 2011). But this changing of the guard, as well as the recovery of certain parts of the pre-1945 historiographical narrative traditions, did not create a context in which economic history thrived. On the contrary, it allowed for a gradual (re)discovery of nationalism as the main ideological and theoretical framework of historical research (Murgescu, 2000, 36–37).

The period from the late 1950s to the late 1960s and early 1970s is regarded as the "liberal" one. Indeed, Romanian historians could with fewer difficulties establish international contacts with their peers or take part in international conferences, such as the international congresses of historical sciences (Stoica, 2012; Velimirovici, 2015a, 91–193). But this short period of time, while important, ended as the Romanian communist regime moved toward the personal rule of Nicolae Ceauşescu, based on his cult of personality, as well as on using the nation as the main propaganda narrative.

Institutionally, the end of the liberalization period is linked to the creation of the Academy of Social and Political Sciences in 1970. This new Academy, directly under the supervision of the Central Committee of the Party, had the role of coordinating all research in social sciences and humanities, including history. All research institutes in these fields, with one exception (the former Party History Institute), were moved from the Romanian Academy to the newly established Academy of Social and Political Sciences in order to be better coordinated politically and ideologically. This political decision was directly connected to the broader phenomena developing in the early 1970s in Romania, where those in political power insisted on asserting their direct control on scientific research (Popa, 2018). In 1975, historical research institutes were once again transferred, this time within the institutional frameworks of universities. This new reform meant that the history research institutes were both under the administrative coordination of the universities and under the ideological control of the Academy of Social and Political Sciences, which retained its role in this regard (Rotaru, 2015, 64–65). The high interests of the Party in strictly regulating and controlling historical outputs can be exemplified by the Central Institute of National History project from 1985. While it was not put into practice, it would have meant the unification of all history institutes into one single institution, directly controlled by the Party apparatus, and a massive reduction in the number of research positions for Romanian historians (Velimirovici, 2015b).

Apart from the academic research institutes, two other institutes played an important role in the Romanian historiographical field. The first one, created initially as the Party History Institute, was renamed in 1966 as the Institute of Historical and Socio-Political Studies, directly under the coordination of the Central Committee of the Party (Zavatti, 2016). This was linked to the Party's (re)discovery of history and nationalism as a hegemonic ideological framework. The other institution, created in 1969, was the Center for Studies and Research on Military History and Theory. Under the significant influence of Ilie Ceaușescu, the brother of Nicolae Ceaușescu, the center quickly became one of the most important actors in promoting a historical narrative based on nationalism (Zavatti, 2016).

As previous research has argued, the three main (sub)periods of the history of Romanian historiography during the communist regime did have an impact on work conditions, especially in relation to internationalization (Iacob, 2011; Stoica, 2012; Velimirovici, 2015a). In the first half of the 1950s, Romanian historians had practically no connection to Western colleagues. From the mid-1950s to the early 1970s, the situation improved. Romanian historians participated in the international congresses of historical sciences without presenting papers in Rome in 1955, but starting in Stockholm in 1960 they had their own contributions, and the participation increased every congress until San Francisco in 1975 (Erdmann, 2005, 313). Beginning with the mid-1970s, the situation regressed again, as the Romanian communist regime decided to return to (neo)Stalinist policies. Thereafter, the level of internationalization, as well as the number of research opportunities abroad, decreased significantly for Romanian historians. Apart from this chronological evolution of working conditions of Romanian historians, there was an institutional difference as well. While data on this topic is still scarce, post-1989 memoirs point out the advantages of historians from the Party History Institute, in comparison with their colleagues from the academic institutes or from universities (Constantiniu, 2007; Stan, 2010).

Diverse Paths Toward Economic History

Andrei Oțetea and the Limits of Romanian Economic History During Communism

One of the leading Romanian historians during the first two phases of the communist rule was Andrei Oțetea (1894–1977) (Hariuc, 2018). He studied at the Sorbonne in Paris from 1919 to 1926, earning his PhD with a thesis on the Florentine historian Francesco Guicciardini (1483–1540). After his return to Romania, he obtained a position at the University in Iași where he taught modern history. Influenced mainly by his economic history professor in Paris, Henri Hauser, Oțetea was one of the few Romanian historians who had already read Marx and who argued for the use of historical materialism and for the paramount importance of the economic factor in historical processes (Oțetea, 1938, 1939, 1941). He also helped persecuted Jews in Iași (Campus, 1994, 639), and in April 1944, together with other main intellectuals, signed a manifesto asking that Romania quit the war against the Allies.

After August 23, 1944, Oțetea entered politics and became a member of the pro-communist Union of Patriots (after 1946 National Popular

Party). In November 1946, he was elected MP and moved to Bucharest, where he was appointed professor of economic history at the university and director of the Institute for World History (now the "N. Iorga" Institute of History). Like other "fellow travelers," he was gradually marginalized after the complete communist takeover. In 1948, Oțetea lost the position of director of the institute and later was purged from the Party but maintained his chair at the university (where the chair of economic history was absorbed into the more encompassing chair of world history).

In the mid-1950s, a group of intellectuals took advantage of the thaw following the death of Stalin and steered a gradual realignment of the communist regime with national cultural traditions. Roller lost his position as vice-president of the Romanian Academy, and in 1955 there was a significant wave of fresh admissions, Andrei Oțetea becoming one of the new members of the Academy. As historiography was one of the main battlegrounds, Oțetea cooperated with other historians trained in the interwar period, especially Constantin Daicoviciu (1898–1973) and Emil Condurachi (1912–1987), in an institutional offensive that led to the demise of the Stalinist group around Mihail Roller (Constantiniu, 2007, 153–234; Papacostea, 1994; Velimirovici, 2015a, 91–147). As part of this change, Oțetea was restored as director of the Institute of History in Bucharest, which he led heavy-handedly from 1956 to 1970; he also became president of the Romanian Academy's section for historical and archaeological sciences. During the late 1950s and the 1960s, Oțetea was instrumental in restoring the relations of Romanian academic institutions and individual historians with their Western counterparts. With the help of the Polish historian Stanislav Schwann, the international connections allowed him to locate several manuscript notes of Marx regarding the Oriental Question and the Romanian Principalities in the 1850s and to prepare a Romanian edition of these texts. Romanian-Polish historiographical relations were established institutionally from the late 1950s, based on a cultural agreement, which provided for the realization of a project entitled Joint Works on the Editing of Polish-Romanian Treatises from the Middle Ages to Today (Hariuc, 2018, 429). Schwann was also a graduate of the University of Cernauti (Chernivtsi), being familiar with the history of the Romanians (Iacob, 2011, 248). The editing of Marx's notes (Marx, 1964), which were very hostile to Russia and generally favorable to the Romanians, served the interests of the Romanian communist leadership, which had distanced itself from the Soviets and tried to legitimate its independent national-communist course (for details and context, see Hariuc, 2018, 427–450). Although his academic authority and institutional position were strong, Oţetea faced the hostility of powerful political and academic actors, such as Constantin Daicoviciu and Miron Constantinescu (Hariuc, 2018, 470–484). In 1970, he had to retire from the leadership of the Institute of History, and health problems marred his last years until he passed away in 1977 in Paris.

Andrei Oţetea was not an economic historian by training. At the Sorbonne, he studied history in the framework of the Faculté des lettres in conjunction with literature, and took courses also at Sciences Po, which was a separated higher education institution. Economic history was taught at that time by Henri Hauser (1866–1946), himself a historian interested in economic issues and persuaded of the importance of the economy, but not trained as an economist. The formation of both Hauser and Oţetea occurred before the mathematization of economics, at a time when "economics was not highly formalised" (Grenier, 2016, 123). Even in the 1930s, when Henri Hauser was one of the founders of the International Scientific Committee on Price History, he argued for the need to base price history on the preserved historical records of transaction prices and against the quantitative use by Ernest Labrousse of series of official prices (Dumoulin, 1990). Influenced by Hauser, by his doctoral supervisor Henri Hauvette, and by the writings of Henri Sée, Oţetea specialized in the study of the Italian Renaissance and studied it as a complex phenomenon, closely connected to the emergence of a modern economy shaped by capitalism. As such, he evolved and remained more of a generalist historian keenly aware of the importance of economic processes.

This pattern of approaching economic history as an indissoluble part of general history can be observed throughout the whole career of Oţetea. It persisted also in his synthesis about the Renaissance and Reformation (Teodor, 1994): in his studies about the Oriental Question, where trade interests were considered a crucial element in shaping the conduct of various states (Cernovodeanu, 1994): and in his studies about the revolution of 1821 and about the national movements of the nineteenth century. Even if in his later career Oţetea acknowledged the new trends in economic history, such as business history (see, e.g., the third edition of his synthesis about the Renaissance and Reformation—Oţetea, 1968, 11) or quantitative analysis (Oţetea, 1977, 8), these acknowledgements were confined to short mentions in prefaces and sometimes to the mere use of the results obtained by others through these approaches and did not lead to a change in his own methods or approach. Even if he stressed the importance of the economy in the general historical evolution, Oţetea did not encourage the emergence of distinct institutional structures of economic history. Although in 1947–48 he held the first (and single) chair of economic history at the University of Bucharest, he agreed into its absorption in the more general chair of (medieval, modern, and contemporary) world history, which he led from 1948 to his retirement from the university in the 1960s.

Of particular significance is Oţetea's stand regarding the second serfdom, which in the 1950s and 1960s was a widely discussed aspect of the debates about the transition to capitalism and the divergence between Western and Eastern Europe. According to his student and collaborator Florin Constantiniu (Constantiniu, 1994, 682), Oţetea came to this issue not by having contacts to the historians who analyzed it in the 1950s, but through his lectures on older works written by Engels and by Henri Sée. At the moment when Romanian historians started to reconnect internationally, Oţetea published a paper on the second serfdom in the Romanian Principalities (1746–1821) in a collective volume prepared for the International Congress of Historical Sciences organized in Rome in 1955 (Oţetea, 1955). In this paper, Oţetea argued that after the demise of serfdom through the reforms enacted by Prince Constantin Mavrocordat in Wallachia (1746) and Moldavia (1749), the landowners gradually increased the exploitation of legally free peasants and established thus a second serfdom. Oţetea's thesis was controversial in Romanian historiography, especially because grain exports were still limited in the eighteenth century and also because the labor obligations of the Wallachian and Moldavian peasants, even increased, were still moderate compared to other parts of Central Europe (e.g., Poland or Transylvania).

Oţetea was stubborn in his search for a second serfdom, and soon after he took the leadership of the Institute of History, he established a research unit which studied the "demise of feudalism" and published two volumes of documents regarding the agrarian relations in the eighteenth century. As one of the young historians included in this research group remembers, in 1957–1958, Oţetea summoned weekly the members of this unit to present the findings of their archival research (Constantiniu, 1994, 682). Several younger members of this research group were assigned topics related to the social and economic history of the eighteenth-nineteenth centuries for their doctoral dissertations. Among his closest collaborators, Şerban Papacostea (1928–2018) wrote a doctoral thesis on Austrian rule in Oltenia (1718–1739), while Florin Constantiniu (1933–2012) was assigned the study of agrarian relations in Wallachia during the eighteenth century. On this basis, Andrei Oțetea refined his theory, included the second serfdom in the more general discussion about the transition from feudalism to capitalism (Oțetea, 1960a), and even accepted that the second serfdom became significant only after the peace treaty of Adrianople (1829) (Oțetea, 1960b).

Oțetea intended to present his findings at the International Congress of Historical Sciences organized in Stockholm in 1960 but his intent was cut short as the general secretary of the International Committee of Historical Sciences Michel François informed his Romanian counterparts that Oțetea's application to present a report on this topic was not included in the program of the congress.[1] Oțetea also failed in his attempt to include his theory about the second serfdom in the Romanian Principalities in the third volume of the official history treatise of the Romanian Academy, where he was editor-in-chief. After an inconclusive debate organized at the Institute of History in April 1960, with several important historians resisting his attempt (Discuția publică…, 729–737), Oțetea gave up his attempt to include the second serfdom as an organizing concept in the history treatise. According to Florin Constantiniu, Oțetea abandoned his idea because of the stubborn opposition of his friend and ally David Prodan (1902–1992), who argued that in Transylvania he could not find any second serfdom (Constantiniu, 2007, 263). Yet, as Andi Mihalache noticed, during the gradual distancing of the Romanian communists from the Soviet hegemony in the early 1960s the ideological context became less favorable for using this Marxist concept, which outlined similarities between Romania and other parts of Central and Eastern Europe (Mihalache, 2003, 167). Although he continued to put passion in this topic even later (Constantiniu, 2007, 264–267), in the 1960s Oțetea invested less in the study of the second serfdom. Although he had access to foreign literature and to travel, he was absorbed by various institutional responsibilities and conflicts and perhaps unable to adjust to the new evolutions of economic history at this stage of his career. Therefore, he did not pay attention to the new contributions

[1] Arhiva Institutul de Istorie "Nicolae Iorga," Fond Comitetul Național al Istoricilor Români, Letter from Michel François, December 29, 1959.

regarding the second serfdom of a younger generation of Polish and Hungarian economic historians, such as Witold Kula and Zsigmond Pál Pach, and his main recent intellectual reference remained Soviet historian Sergei Danilovich Skazkin, whose synthetical article in Voprosy Istorii was translated in the same year in Romanian (Skazkin, 1958).

Oţetea's ambivalence regarding economic history is highlighted also by his relationships with his students and subordinates. He encouraged many of them to engage in research on themes of social and economic history, and some of his younger colleagues from the University of Bucharest or from the Institute of History produced doctoral theses which were milestones in Romanian economic history. For example, Radu Manolescu (1929–2005) studied the customs records of Braşov during the sixteenth century (Manolescu, 1965), Sergiu Columbeanu (1924–1978) focused on the economy of eighteenth-century Wallachian domains (Columbeanu, 1974), Georgeta Penelea (1938–) on the trade fairs in Wallachia from 1774 to 1848 (Penelea, 1973), Mircea Popa (1940–1987) on the history of money and prices in Wallachia from 1774 to 1831 (Popa, 1978) and Constantin Buşe (1939–2019) on the international trade through Galatz from 1837 to 1883 (Buşe, 1976). Yet, Oţetea also supervised doctoral theses that had no connection with economic history, and most of his disciples who started with research on economic history shifted later to other fields of interest, such as social, political, or cultural history.

Oţetea's case highlights the importance of timing in establishing international academic contacts. Having studied in Paris during the early 1920s, he was influenced by the rather soft economic history championed by Henri Hauser and did not connect with the quantitative approach championed by Ernest Labrousse in the 1930s (Grenier, 2016, 115), although he was vaguely acquainted with him and Labrousse's works on the eighteenth century would have appealed to him. Focused on Romanian issues, Oţetea also missed the workings of the International Scientific Committee on Price History and the impact of the emerging Annales school. After 1945, Oţetea had almost no opportunities to continue contacts with Western scholarship, which meant that he skipped the impact of Braudel's *La Méditerranée et le monde méditerranéen*..., as well as the effervescent evolution of economic history during the 1950s. When he was able to reconnect starting with the late 1950s, Oţetea was simply too embedded in his own approach and therefore unable to renew himself.

At the same time, the Romanian institutional landscape did not favor specialization in economic history. Traveling to international congresses depended heavily on being part of official delegations, and the international congresses of historical sciences (Stockholm 1960, Vienna 1965, Moscow 1970) overshadowed the emerging specialized international congresses of economic history. As Oțetea had no real connections to the historians who established the IEHA, he did not attempt to take part in any of the economic history congresses and preferred to invest in building up bilateral connections with various French, Italian, German, or American colleagues.

The Romanian National Commission for UNESCO was an important institution in helping Romanian scholars to connect to their colleagues abroad, and after serving as president of the section for social sciences, Oțetea was appointed in 1966 president of the Romanian National Commission and served also as member of the Executive Committee of UNESCO; this allowed him to travel frequently abroad, but also absorbed a lot of his energy in the late 1960s (Hariuc, 2018, 485–496). Oțetea was also part in the establishment of the Association Internationale d'Etudes du Sud-Est Européen as a regional organization supported by UNESCO. In the 1950s and early 1960s the Balkans were more divided than ever. Albania, Bulgaria, Romania, and Yugoslavia had turned socialist at the end of World War II, but divergences among them were huge and adversities grew in time. On the other side of the Iron Curtain, Greece and Turkey were both members of NATO, but their bilateral relations were dismal. In this complicated context, both politicians and academics sought to overcome the divisions and promote academic cooperation as a "proxy for the political cooperation" (Iacob, 2015, 28). The Association Internationale d'Etudes du Sud-Est Européen was established in 1962 as a multilateral organization that brought together not only national member organizations from the countries in the region, but also academic bodies from several other countries, both East and West. While the General Secretariat functioned in Bucharest, the board was multinational and its congresses were organized by rotation (Sofia 1966, Athens 1970, Bucharest 1974, Ankara 1979, Belgrade 1984, and so on). In the multidisciplinary setting of the Association Internationale d'Etudes du Sud-Est Européen, economic history was only a small part, and its main promoter was no Romanian, but the Bulgarian Nikolai Todorov.

Gheorghe Zane, an Economist Turned Economic Historian

Gheorghe Zane (1897–1978) presents a different case. He became first an associate professor and later a professor at the University of Iaşi, where he taught political economy, the history of economic doctrines, and finance. His main research focused during this period on the economic history of the Romanian Principalities in the late eighteenth and nineteenth centuries, with remarkable publications on economic and juridical uncertainty (Zane, 1925), on international trade (Zane, 1926, 1927a), on monetary problems (Zane, 1927b), and on the evolution of the exchange economy (Zane, 1930). He also entered politics, becoming a member of the Peasant Party (1925) and later of the National Peasant Party. He was one of the main theorists of the peasant state and was elected deputy in the Romanian Parliament in several legislatures: 1928–31, 1932–33, and 1946–47. This political engagement cost him during the communist takeover. He was arrested in 1947, released and briefly hired at the Library of the Academy and at the History Institute in Bucharest, then arrested again in 1952, and released in 1953.

Zane's former assistant Alexandru Bârlădeanu (1911–1997), who had become a leading member of the Party, helped him get hired at the Institute of Economic Research of the Romanian Academy, going through various positions, starting as researcher and becoming head of department and a senior member of the scientific council. At the institute he capitalized on his expertise in the history of nineteenth-century revolutionary thought—he had started already in 1940 to edit the writings of Nicolae Bălcescu, one of the 1848 revolutionaries who continued to be praised also during the communist regime—and continued editing the works and letters of Bălcescu. Zane also coordinated a collective edition of nineteenth-century Romanian economic writings (Zane, 1960) and helped Oţetea in editing the manuscripts of Marx (Hariuc, 2018, 433). Later, he focused on the development of Romanian industry in the nineteenth century (Zane, 1970a) and continued his studies on the economic history of the late eighteenth and nineteenth centuries (see Zane, 1980).

During the partial liberalization of the 1960s, Zane was for a time unable to take advantage of the emerging opportunities for international contacts. With the help of Alexandru Bârlădeanu, he became in 1965 corresponding member of the Romanian Academy (full member in 1974). He also capitalized on contacts with Western scholars who visited the Institute of Economic Research (Zaman et al., 2013, 21–22), started to publish in French, and finally succeeded in establishing contacts in France, where he published a paper on the influence of Saint-Simon and Fourier in the Romanian Principalities (Zane, 1970b) and became member of the Association "Tiers Monde" and of the editorial board of the journal *Mondes en developpement*. Zane also took advantage of the ascent of Miron Constantinescu (1917–1974) as head of the Academy of Social and Political Sciences (March 1970), which centralized control of a large part of the Romanian research institutions, including the Institute of Economic Research (Popa, 2018). He became a member of the economics section of the Academy of Social and Political Sciences and was included in the Romanian delegation sent to the 5th International Economic History Congress organized in Leningrad in August 1970. He seconded Constantinescu in the establishment of the Committee for Economic History and History of Economic Thought, which was supposed to be the Romanian institutional member of the IEHA and became the first president of this committee. Zane obtained approval to edit a monthly academic journal—*Studii de istorie a gândirii economice* [*Studies in the History of Economic Thought*], renamed in 1972 *Memoria Oeconomica*. In a small format, he managed to publish 29 issues, offering thus a platform for scholars who linked current economic issues with pre-communist economic thought and with contemporary foreign contributions to economics. Despite being helped by a younger colleague at the institute, Gheorghe Dobre, Zane was unable to continue publication after 1973, when the Institute of Economic Research was reorganized and its coordination was taken over by a different body, the Supreme Council of Economic and Social Development (Zaman et al., 2013, 29–31).

Excepting a short period starting in the mid-1960s and continuing into the early 1970s, Zane's contacts with foreign economic historians were limited. Although his expertise in nineteenth-century economic thought and in industrialization was suited to foster meaningful interaction, at the moment he managed to establish contacts, Zane was too late in his career to be able to benefit fully from these connections.

Romanians at the International Economic History Congresses

The International Economic History Congresses started to be organized in 1960, the first one (labeled as "conference") occurring in Stockholm just before the International Congress of Historical Sciences (Berg, 2015). Apparently, no participant from Romania was present at the economic history reunion in Stockholm nor at the congresses held in Aix-en-Provence (1962), München (1965), or Bloomington (1968). The first time Romanians participated at an IEHA event was the 5th International Economic History Congress organized in Leningrad in August 1970.

There are several explanations for this late start. First, in communist Romania, even during the more liberal 1960s, traveling abroad was still a controlled privilege, and obtaining approval for travel to a socialist country was easier (and often cheaper) than for travel to a Western country. Besides, the IEHA event in Leningrad was organized just a week before the 12th International Congress of Historical Sciences held in Moscow from August 16 to August 23, 1970. Second, the Academy of Social and Political Sciences, led by the influential Miron Constantinescu with the aim of centralizing research and also international contacts, was founded in March 1970 (Popa, 2018). Constantinescu, who had been a leading political figure during the communist regime and was again an important star after having been marginalized in the late 1950s and the first part of the 1960s, was a sociologist by training who had ambitions to prove himself as a leading intellectual figure in all the social sciences, including economics and history. Constantinescu seized the opportunity to lead a delegation of Romanian economic historians to the Leningrad IEHA conference before joining the larger group of historians at the Moscow Congress of Historical Sciences.

Romania was represented at the Leningrad congress by eleven economic historians.[2] The list of participants highlights the complicated relations between historians and economists in communist Romania. Besides Miron Constantinescu the delegation included seven economists: Gheorghe Zane, N. N. Constantinescu (1920–2000), Victor Axenciuc (1928–2014), Gheorghe Dobre, Vasile Pilat, Aneta Spiridon, Gheorghe Surpat; and three historians: Dionisie M. Pippidi (1905–1993), Samuel Goldenberg (1920–1995) and Paul Cernovodeanu (1927–2006). Among the historians, Dionisie Pippidi was a distinguished classicist and director of the "Vasile Pârvan" Institute of Archeology but had little to do with economic history, while Samuel Goldenberg and Paul Cernovodeanu secured individual invitations to present papers in the section organized by Hermann Kellenbenz on Continental Trade Relations between Eastern and Western Europe before the Railroads and were

[2] Arhivele Naţionale Istorice Centrale, Fond Academia de Ştiinţe Sociale şi Politice. Secţia de Istorie şi Arheologie, 17/1970, 45.

therefore accepted also by Miron Constantinescu. According to post-congress reports, Romanian participants had fourteen short interventions during the congress and presented six papers. Miron Constantinescu's three presentations focused on economic planning methods in socialist Romania, the social and economic conditions of the 1848–49 revolution, and the creation of a national economy in Romania. Gheorghe Zane analyzed in his paper the general characteristics of industrial development in Romania during the second half of the nineteenth century, while Samuel Goldenberg and Paul Cernovodeanu focused on the economic exchanges between the Romanian Principalities and the West during the sixteenth century and the first half of the seventeenth century, and between 1650 and 1829, respectively.

The congress publications included the papers delivered by Samuel Goldenberg and Paul Cernovodeanu, as well as two interventions of Miron Constantinescu (van der Wee et al., 1974–1976, II, 128–133; VI, 195–214; VII, 27–30); the paper prepared by Gheorghe Zane had been published separately in Bucharest in preparation for Leningrad and probably distributed there (Zane, 1970c). During the Leningrad congress, Romania was unanimously accepted as member of the International Association of Economic History and presented its intention to host future international congresses (Studii de istorie și gândire economică, 1971, 115–116). Apparently, this meant a bid to host the 1978 international economic history congress (Berindei & Lungu, 1970, 1235), but even the intention to obtain such a responsibility immediately after joining the organization and without having a representative in the Executive Committee showed how little Miron Constantinescu and his followers understood about the real workings of the IEHA.

This failed bid highlights a more general feature of the international involvement of Romanian intellectuals. As the permission to participate in international conferences was considered a privilege awarded by the regime, some scholars tried both to strengthen their position in various international organizations and networks and to convince the Romanian authorities that hosting prestigious academic conferences would enhance the country's international reputation. This was true not only for the failed attempt to host an IEHA congress, but also for the successful bid to host the 1980 International Congress of Historical Sciences; in assessing the outcome, we must consider that Romanian historians had participated in congresses of historical sciences without interruption since 1955 (besides being active in the interwar period). In contrast, the Romanian bid to host the 1978 IEHA congress fell into oblivion, and the congress was organized in Edinburgh.

Although soon after the Leningrad congress Miron Constantinescu chaired the official establishment of the Romanian Committee for Economic History and History of Economic Thought, which was to be the institutional Romanian member of IEHA, the Romanian participation in IEHA activities was largely discontinued. In fact, at the 6th International Economic History Congress held in Copenhagen in 1974 only agrarian historian Eugen Mewes, who presented a very general paper on the notion of "economic history," participated (Mewes, 1981, 31–41).

As the Romanian communist regime evolved into an autarchic and nationalistic state, the opportunities to participate in international scientific events decreased significantly for Romanian scholars. Our research did not find any Romanian participation at the Edinburgh 1978 or Bern 1986 Economic History Congresses.

The only exception from the late 1970s and 1980s was the 8th International Economic History Congress held in Budapest in 1982. Victor Axenciuc, an economic historian working at the Institute for Socialist Economy, who had participated in the Leningrad congress in 1970, successfully proposed a C-session on the national economies in south-eastern Europe during the interwar period, but he was not able to attend the event, supposedly because of health problems (Bozga, 1983, 83). His session included two other Romanian economic historians teaching at Academy of Economic Studies, Ivanciu Nicolae-Văleanu (1927–), professor of history of economic doctrines, who took over the moderation of the session and briefly presented Axenciuc's paper, and economic historian Vasile Bozga (1930–2017). Another Romanian economic historian, Constanța Bogdan, had two interventions in different sessions. Bogdan's intervention regarding the Romanian public debt 1864–1940 was appreciated positively by prof. N. Petrović, the Yugoslav chair of the session. Besides presenting papers and contributing to the discussions in various sessions, the three Romanian participants distributed papers printed in Bucharest, which contained their own texts and the contributions of other ten economists who did not make it to participate in person to the congress in Budapest (Bozga, 1983, 84–85).

Conclusion

Communist rule had a significant impact on Romanian economic history, but this impact was not homogeneous over time. In a first phase, Stalinism included purges in universities and sometimes also outright repression against intellectuals including some of the few economic historians (see the case of Zane above), while the East-West divide almost completely blocked academic interactions across the Iron Curtain. Then partial liberalization during the late 1950s and 1960s included some openings toward the West, as well as renewed career opportunities for intellectuals who had been marginalized during Stalinism. After 1971, as Ceaușescu moved toward a more self-centered pattern of national communism (Verdery, 1991), the upsurge of contacts with the West slowed down, and finally in the 1980s went negative. While these variations were important, we have to consider that throughout the whole period of communist rule, traveling abroad and even cultivating contact with foreigners were considered privileges bestowed by the regime, which could be terminated at will by the party or by the Securitate; besides, as outlined by Bogdan Iacob, the regime preferred to send to the West established scientists rather than young experts (Iacob, 2011, 414).

In communist Romania, economic history suffered more from these political vicissitudes also because of its constitutive weaknesses. As explained in this chapter, economic history did not form a unified autonomous professional field. Historians who wrote about topics of economic history were not specialized in this field and acted more like generalists in the field of history. This is true not only in the case of Andrei Oțetea, but also among his slightly younger colleagues who had established first international contacts in the interwar period and cooperated in actively promoting the opening toward the West, like Emil Condurachi and Mihai Berza (1907–1978), and among their disciples educated after 1945. Besides, after lip-service paid to Marxism in the 1950s, most Romanian historians returned to the pre-communist habitus of being reluctant to apply general theories (Oțetea was an exception in this respect) and preferred to take refuge in document-based reconstructions of concrete historical facts. Simultaneously, the economic sciences developed into a scholastic endeavor legitimizing the policies of the regime, and fact-based empirical research was rare in the economic profession. While the use of modeling and mathematical methods was confined to a small sector linked to planning and cybernetics (Cojanu, 2021), some of the more open-minded economists took refuge in the history of economic thought, which was relatively prestigious in the general context of not-inspiring economics.

In time, distances grew because in many other countries the field of economic history evolved. The more traditional patterns of economic history were supplemented (but not yet replaced) by approaches which were closer to economic theory, more formalized and using quantitative data and mathematical models. Romanian economic historians, few as they were, did not share these evolutions. The generational structure was also part of the problem. Economic historians educated trained in the interwar period (e.g., Oțetea and Zane, but also others), could benefit from the openings of the late 1950s and 1960s late in their careers, and passed away in the 1970s without being replaced at par by a next generation educated after 1945. Compared with their Hungarian counterparts (Berend, 2009), Romanian economic historians educated after 1945 did not have the social capital and the opportunities to bond with their Western colleagues in the 1970s and 1980s and were bound to a more national setting in which economic history was not a major rallying point.

References

Arhiva Institutul de Istorie „Nicolae Iorga," Fond Comitetul Național al Istoricilor Români.

Arhivele Naționale Istorice Centrale, Fond Academia de Științe Sociale și Politices.

Berend, I. T. (2009). *History in My Life: A Memoir in Three Areas*. Central European University Press.

Berg, M. (2015). East-West Dialogues: Economic Historians, the Cold War, and Détente. *The Journal of Modern History, 87*, 36–71.

Berindei, D., & Lungu, T. (1970). Cel de-al XIII-lea Congres internțional de științe istorice și contribuția delegației române la lucrări sale. *Studii. Revistă de istorie, 23*(6), 1229–1235.

Bosomitu, Ș., & Burcea, M. (2012). Mihai Roller (1908–1958). O scurtă biografie. In *Identitate, social și cotidian în România comunistă, Anuarul Institutului de Investigare a Crimelor Comunismului și Memoria Exilului Românesc* (Vol. VII, pp. 19–65). Polirom.

Bozga, V. (1983). Al VIII-lea Congres internațional de istorie economică. *Revista de istorie, 36*(1), 83–85.

Buşe, C. (1976). *Comerţul exterior prin Galaţi sub regimul de port franc (1837–1883)*. Editura Academiei.

Campus, E. (1994). Andrei Oţetea: rigoare morală şi ştiiţifică în învăţământ şi cercetare. *Revista Istorică, V*(7–8), 639–644.

Catalan, G. (2009). Istoriografia română sub impactul modelului sovietic 1947–1955. In *Intelectualii şi regimul comunist: istoria unei relaţii, Anuarul Institutului de Investigare a Crimelor Comunismului şi Memoria Exilului Românesc* (Vol. IV, pp. 22–25). Polirom.

Cernovodeanu, P. (1994). Andrei Oţetea istoric al chestiunii orientale. *Revista Istorică, V*(7–8), 657–670.

Cojanu, V. (2021). *Repere ale politicii ştiinţei în România*. Perspectiva unui economist, Casa Cărţii de Ştiinţă.

Columbeanu, S. (1974). *Grandes exploitations domanioles en Valachie au XVIIIe siècle*. Editura Academiei.

Constantiniu, F. (1994). Andrei Oţetea şi a doua iobăgie: o dezbatere inutilă? *Revista Istorică, V*(7–8), 681–688.

Constantiniu, F. (2007). *De la Răutu şi Roller la Muşat şi Ardeleanu*. Editura Enciclopedică.

Discuţie publică cu privire la „Iobăgia a doua în ţările din centrul şi răsăritul Europei" şi „Iobăgia a doua în Moldova şi Ţara Românească". (1961). *Studii. Revistă de istorie* (3–4), 729–737.

Dumoulin, O. (1990). Aux origines de l'histoire des prix. *Annales. Économies, sociétés, civilisations, 45*(2), 507–522.

Enciclopedia istoriografiei româneşti. (1978). Editura Ştiinţifică şi Enciclopedică.

Erdmann, K. D. (2005). *Toward a Global Community of Historians. The International Historical Congresses and the International Committee of Historical Sciences, 1898–2000* (Edited by J. Kocka & W. J. Mommsen in collaboration with A. Blänsdorf, Translated from the German by A. Nothnagle). Berghahn Books.

Grenier, J.-Y. (2016). Economic History in France: A Sonderweg? In F. Boldizzoni & P. Hudson (Eds.), *Routledge Handbook of Global Economic History* (pp. 113–129). Routledge.

Hariuc, M. (2018). *Andrei Oţetea – profil intelectual şi activitate istoriografică* (PhD thesis, „Alexandru Ioan Cuza" University of Iaşi). https://rei.gov.ro/teze-doctorat

Iacob, B. C. (2011). *Stalinism, Historians, and the Nation: History-Production Under Communism in Romania (1955–1966)*. A Dissertation in History, Budapest, Hungary. https://www.etd.ceu.edu/2011/hphgab02.pdf

Iacob, B. C. (2015). Southeast European Studies During the Cold War: Aspects of International Institutionalization (1960S–1970s). In *New Europe College Stefan Odobleja Program Yearbook, 2014–2015* (pp. 21–52). New Europe College.

Manifestări ştiinţifice. (1971). *Studii de istorie şi gândire economică, XIII.*

Manolescu, R. (1965). *Evoluţia schimbului de mărfuri al Ţării Româneşti şi Moldovei cu Braşovul în secolele XIV–XVI.* Editura Ştiinţifică.

Marx, K. (1964). *Însemnări despre români: manuscrise inedite, publicate de A. Oţetea şi S. Schwann.* Editura Academiei.

Mewes, E. (1981). On the Contents of the Notion of "Economic History". *Terra nostra,* IV, 37–41.

Mihalache, A. (2003). *Istorie şi practici discursive în România „democrat-populară".* Albatros.

Murgescu, B. (2000). *A fi istoric în anul 2000.* All Educaţional.

Opriş, I. (2006). *Istoricii şi Securitatea* (Vol. II). Editura Enciclopedică.

Oţetea, A. (1938). *Concepţia materialistă a istoriei ca metodă de cercetare şi de expunere.* Institutul de Arte Grafice «Brawo».

Oţetea, A. (1939). Cauzele economice ale Renaşterii. *Viaţa Românească,* anul XXXI (1), 40–57.

Oţetea, A. (1941). *Renaşterea şi Reforma, Bucureşti.* Fundaţia Regală pentru Literatură şi Artă.

Oţetea, A. (1955). Le second aservissement des paysans roumains (1746–1821). *Nouvelles études d'histoire, I,* 299–312.

Oţetea, A. (1960a). Consideraţii asupra trecerii de la feudalism la capitalism în Moldova şi Ţara Românească. *Studii şi cercetări de istorie medie, IV,* 307–390.

Oţetea, A. (1960b). Le second servage dans les Principautés danubiennes (1831–1864). *Nouvelles études d'histoire, II,* 325–346.

Oţetea, A. (1968). *Renaşterea şi reforma.* Editura Ştiinţifică.

Oţetea, A. (1977). *Pătrunderea comerţului românesc în circuitul internaţional (în perioada de trecere de la feudalism la capitalism).* Editura Academiei.

Papacostea, Ş. (1994). Andrei Oţetea, Director al Institutului de Istorie „Nicolae Iorga". *Revista Istorică, V*(7–8), 629–638.

Penelea, G. I. (1973). *Les foires de la Valachie pendant la période 1774–1848,* Editions de l'Académie de la République Socialiste de Roumanie.

Popa, C. (2018). *Intelectualii lui Ceauşescu şi Academia de Ştiinţe Sociale şi Politice (1970–1989).* Litera.

Popa, M. (1978). *La circulation monetaire et l'evolution des prix en Valachie (1774–1831).* A.I.E.S.E.E.

Rotaru, M. (2015). *Istoria Universităţii din Bucureşti în anii 1980* (PhD thesis). University of Bucharest.

Skazkin, S. D. (1958). Osnovnye problemy tak nazyvaemogo «vtorogo izdanija krepostnichestva» v Srednej i Vostochnoj Evrope. *Voprosy Istorii, 2,* 96–119.

Stan, A. (2010). *Istorie şi politică în România comunistă.* Curtea Veche.

Stoica, S. (2012). *Istoriografia românească între imperativele ideologice şi rigorile profesionale 1953–1965.* Meronia.

Teodor, P. (1994). Andrei Oţetea istoric al Renaşterii. *Revista Istorică, V*(7–8), 645–653.

van der Wee, H., Vinogradov, V. A., & Kotovsky, G. G. (Eds.). (1974–1976). *Fifth International Congress of Economic History, Leningrad 1970* (Papers Vols. I–VIII). Moskva.

Velimirovici, F. (2015a). *Istorie şi istorici în România comunistă (1948–1989)*. Mega.

Velimirovici, F. (2015b). The Evolutions of Romanian 'Historians' Front' During Stalinism and Late Socialism. RCP's Plans for Establishing a Central Institute for National History. *Studia Universitatis Babes-Bolyai – Historia, 60*(2), 1–28.

Verdery, K. (1991). *National Ideology Under Socialism: Identity and Cultural Politics in Ceausescu's Romania*. University of California Press.

Zaman, G., Vasile, V., & Georgescu, G. (2013). *Institutul de economie naţională. 60 de ani de cercetare ştiinţifică*.

Zane, G. (1925). *Impreviziunea. Cîteva cercetări asupra dezvoltării economice şi juridice a Ţărilor Române la începutul secolului XIX*. Institutul Modern de Arte Grafice N. V. Stefaniu si D. Staierman.

Zane, G. (1926). *Un veac de luptă pentru cucerirea pieţei româneşti*. Iaşi.

Zane, G. (1927a). Die österreichischen und die deutschen Wirtschaftsbeziehungen zu den rumänischen Fürstentümern 1774–1874. *Weltwirtschaftliches Archiv, 26*, 30–47.

Zane, G. (1927b). *Probleme monetare din trecutul economiei noastre naţionale*.

Zane, G. (1930). *Economia de schimb în Principatele Române*. Editura Casei Şcoalelor.

Zane, G. (1960). *Texte din literatura economică în România–secolul XIX. Bucureşti*, Editura Academiei.

Zane, G. (1970a). *Industria din România în a doua jumătate a secolului al XIX-lea*. Editura Academiei.

Zane, G. (1970b). Le Saint-Simonisme et le Fouriérisme en Roumanie. In *Economies et sociétés*. Cahiers de l'I.S.E.A, IV.

Zane, G. (1970c). *Caractères généraux du développement de l'industrie roumaine au cours de la seconde moitié du XIXe siècle*. Centre d'information et de documentation pour les sciences sociales et politiques.

Zane, G. (1980). *Studii*. Ed. Eminescu.

Zavatti, F. (2016). *Writing History in a Propaganda Institute. Political Power and Network Dynamics in Communist Romania*. Elanders, Södertörn University.

CHAPTER 9

Epilogue

Antonie Doležalová

With the fall of the Iron Curtain, the necessity for academics to dodge between the Scylla of ideological stiffness and Charybdis of intellectual lethargy suddenly disappeared. The years after 1989 full of promises promised also to finally break all the barriers, both personal and ideological, that stood in the way of normal academic cooperation. It seemed that Postan's and Braudel's dream, imprinted into the establishment of IEHA, to help develop and strengthen East-West cooperation (Berg, 2015) could be fulfilled and crowned by success. Were these promises fulfilled? Is the cooperation between countries that were once divided by the Iron Curtain free of barriers, barricades, and biases? In other words, in what measure is the socialist legacy still visible in individual national post-socialist historiographies, and in what measure has Western academia managed to be inclusive and overcome the image of Central and Eastern European scholars as relics of manipulated and centrally controlled socialist academies?

A. Doležalová (✉)
Institute of History, Czech Academy of Sciences and Faculty of Social Sciences, Charles University, Prague, Czech Republic
e-mail: ad875@cam.ac.uk

A. Dolezalova and C. Albrecht (eds.), *Behind the Iron Curtain*, Palgrave Studies in Economic History,
https://doi.org/10.1007/978-3-031-31578-7_9

From the geopolitical and national perspective, the ostensible unity of historiographies in three countries, the Soviet Union, Czechoslovakia, and Yugoslavia, fell apart within a few years with the political division of these countries. By contrast, East German historiography was unified with its Western counterpart, giving birth to a single German historiography. Other profound changes that ended totalitarian regimes and exposed not only policies and societies, but also academies, to the rest of the world touched all of the historiographies under discussion despite their different traditions, organization of science, funding distribution, or commitment to Marxism.

After 1989, economic history, which had been a dominant stream of historiography only in the GDR and Poland before 1989, was moved to the periphery in all countries. This was in part because economic history as a history of workers' movement, jointly with philosophy, political economy, and scientific communism, was at all universities part of the institutes of Marxism-Leninism that were dissolved. The former members either left academia or retrained themselves in more convenient fields of expertise such as microeconomics, macroeconomics, or political science; some of them found new positions in the departments of history at newly created, state or private, often regional, universities as the majority of these new universities focused on the humanities and social sciences. In addition, historians refocused their research on more attractive and previously prohibited areas of political, cultural, and social history.

The complex process had its human costs. Not only professional criteria but often political, personal, and even ideological maneuvering dictated again the reconstruction of academia and its adaptation to new conditions. Contradictory aspects of implementing Western criteria of evaluating scientific performance are evident in East German historiography. Approximately, three-quarters of all East German historians lost their jobs after unification (Fair-Schulz & Kessler, 2017, 3). Kuczynski's Institute for Economic History at Academy of Sciences, the world's largest workplace in the field, as well as Department of Economic History at the University of Economics in Berlin, were closed down; fewer than a handful of former members of Kuczynski's institute were able to find new positions in German academia. Former East German academicians were replaced by newly appointed Western professors (Fair-Schulz, 2017).

Even though there was no Western counterpart in other post-communist countries, economic history followed very similar trajectories across the region. Only a limited number of pupils continued in what had been successful, internationally recognized schools in economic history in Poland or Hungary; in Hungary, the discipline was even more weakened by the untimely death of György Ránki and by Iván Berend's emigration to the United States. The Department of Economic History at the Corvinus University in Budapest was dissolved. In Czechoslovakia, the Department of Economic History at University of Economics in Bratislava was also dissolved. On the other hand, new workplaces were established at the Faculty of Arts of Charles University in Prague, at Eötvös Loran University in Budapest, and at regional universities in Russia or Czechoslovakia; with, however, only limited numbers of students. More Ukrainian economic historians can be found at universities in Canada and the United States than in Ukraine.

Some journals, such as the major Polish journal of economic history *Roczniki Dziejów Społecznych i Gospodarczych*, which was established in 1931 in Lviv, left the scene; others, as the Czechoslovak and then Czech *Hospodářské dějiny/Economic History*, lost their periodicity and prestige. The East German *Jahrbuch für Wirtschaftsgeschichte* was stopped and moved to Frankfurt/Main. On the other hand, a few new journals were established, but with the exception of the Russian *Journal of Globalization Studies* these new journals did not gain international recognition despite being published in the English language.

Research in economic history lost its former systemic nature and active individuals were even more marginalized. As economic history in the Eastern bloc—with the exception of Poland—was a matter of topics, the adaptation to the new political, social, and research conditions after 1989 was accompanied by the large escape from research on the medieval and early modern eras to modern and contemporary history. Even more, while in the communist era the older historical periods were a kind of harbor that allowed historians to avoid ideological pressure, focus on topics other than economic history improves scholars' chances to obtain financial support today. Regardless of the new openness of western borders, the refocusing of economic history was accompanied by an escape from general to national history.

Despite the fact that before 1989 only those economic historians who treated the discipline in a holistic way, saw history in its unity, and openly used Karl Marx's historical materialism succeeded on the international scene, with only a few exceptions, after 1989 economic historians defined themselves as non-Marxists. Marxism as a methodology was marginalized and former Marxists exposed to public rejection as anti-communist sentiment equated communism to Marxism; paradoxically enough, former communist nomenklatura adapted to the new conditions and positioned themselves in the state apparatus and newly established private businesses. Marxism and any type of research critical of capitalism were ostracized and eliminated in the same oppressive way that these approaches had been prioritized during the communist era. Moreover, methodological chaos entered the scene accompanied by fascination with the literary turn, postmodernism, and cultural and political history. Historians are not trained to use economic tools of research in any country of the former Soviet bloc.

Not only is substantial influence from international developments missing in Central and Eastern European historiographies today. Economic historians from post-communist countries are not visible on the international scene either. After 1990 the representatives of East European countries lost the significance they had once held in meetings of the IEHA. Just before his death, Ránki was elected vice president of the International Committee of Historical Sciences, a position that was inherited by Berend, who finally became its president in 1995–2000 (Kövér, 2016, 253, note 29). For a short time after 1990, travel for younger scholars from the former Eastern bloc was funded (Berg, 2015, 65). This was why in 1990, at Leuven's World Economic History Congress, eight of 25 dissertations presented in competition were from the former Eastern bloc; that number has never been exceeded since that time. At the same congress, in the bloc of debates and controversies, Hungary organized one of seven panels; in the bloc of major topics and current research, the Soviet Union and the GDR jointly organized one of sixteen panels; and in the bloc of workshops on more specialized topics former socialists historians organized one of 44 events. At the World Economic History Congress in Paris in 2022, only four out of 200 accepted panels were organized by scholars from the former Eastern bloc.

Only Russian academia was able to extend bilateral cooperation with the West, often supported by western private foundations (Borodkin, 2016, 209). In other countries, the exile community helped to open academia to the world by inviting scholars to participate in international projects. Alice Teichova played a significant role in this regards. After 1989, she opened the doors of the international scientific community to an entire generation of Central and Eastern European historians by inviting them to projects, conferences, and publications she organized.

What I use to call the "Alice Teichova factor" had, surprisingly, a negative effect in the following decade. When Alice Teichova completed her last project a few years after the beginning of the new millennium, nobody from the region replaced her in creating multinational projects or in incorporating post-communist historiographies into the mainstream. In all countries, economic history has been closing to the world by the emphasis on national history with the simultaneous vanishing of general, comparative, or global history.

The long and cold eastern shadow is still an inherent part of individual national historiographies in the countries that once stood behind the Iron Curtain. What once had the potential to promote dialogue between East and West has gone unfulfilled. To his last days, Mikuláš Teich used to say with a smile on his lips that when the scientific community strives for truth, it must function as a communist society: cooperate and share its findings.

References

Berg, M. (2015). East-West Dialogues: Economic Historians, the Cold War, and Détente. *Journal of Modern History, 87*(1), 36–71.

Borodkin, L. (2016). Economic History from the Russian Empire to the Russian Federation. In P. Hudson & F. Boldizzoni (Eds.), *Routledge Handbook of Global Economic History* (pp. 195–213). Routledge.

Fair-Schulz, A. (2017). The Dissolution of the Institute for Economic History at the Academy of Sciences. In A. Fair-Schulz & M. Kessler (Eds.), *East German Historians Since Reunification* (pp. 155–164). SUNY Press.

Fair-Schulz, A., & Kessler, M. (Eds.). (2017). *East German Historians Since Reunification*. SUNY Press.

Kövér, G. (2016). Crossroads and Turns in Hungarian Economic History. In P. Hudson & F. Boldizzoni (Eds.), *Routledge Handbook of Global Economic History* (pp. 242–257). Routledge.

Tenth International Economic History Congress, Leuven 1990. Programme.

van der Wee, H., & Aerts, E. (Eds.). (1990). *Debates and Controversies in Economic History*. Leuven University Press.

Index

A. Dolezalova and C. Albrecht (eds.), *Behind the Iron Curtain*, Palgrave Studies in Economic History,
https://doi.org/10.1007/978-3-031-31578-7

SPRINGER NATURE

GPSR Compliance

The European Union's (EU) General Product Safety Regulation (GPSR) is a set of rules that requires consumer products to be safe and our obligations to ensure this.

If you have any concerns about our products, you can contact us on ProductSafety@springernature.com

In case Publisher is established outside the EU, the EU authorized representative is:

Springer Nature Customer Service Center GmbH
Europaplatz 3
69115 Heidelberg, Germany

The manufacturer's authorised representative in the EU is Springer Nature Customer Service Centre GmbH, Europaplatz 3, 69115 Heidelberg, Germany. If you have any concerns regarding our products, please contact ProductSafety@springernature.com

Printed and bound by CPI Group (UK) Ltd, Croydon, CR0 4YY
30/09/2026
02225494-0004